R. I. Orazbayeva
D. N. Kadyrova
S. S. Rakhimkhodjaev

Designing of costume fabrics and their technology

R. I. Orazbayeva
D. N. Kadyrova
S. S. Rakhimkhodjaev

Designing of costume fabrics and their technology

ScienciaScripts

Imprint

Cover image: www.ingimage.com

This book is a translation from the original published under ISBN 978-620-7-48431-7.

Publisher:
Sciencia Scripts
is a trademark of
Dodo Books Indian Ocean Ltd. and OmniScriptum S.R.L publishing group

120 High Road, East Finchley, London, N2 9ED, United Kingdom
Str. Armeneasca 28/1, office 1, Chisinau MD-2012, Republic of Moldova, Europe
Managing Directors: Ieva Konstantinova, Victoria Ursu
info@omniscriptum.com

Printed at: see last page
ISBN: 978-620-8-36993-4

Contents

Annotation

*The work is devoted to the design with specified properties and production technology of suit fabrics. The criterion of fabric structure assessment is air permeability, which determines quantitative and qualitative assessment of fabric structure. The technique of designing of clothing (costume) fabrics according to the given air permeability is developed, in particular the air permeability coefficient, the percentage of unfilled and filled with fibrous material area of fabric, the average yarn number (linear density), the fabric density and the equation of air permeability for the given fabric are determined. The difference between the experimental values of air permeability and the calculated values of the equation for a given fabric is 16%, which is due to the weave of the fabric, composition, type and twist of the yarn. The equation of the air permeability coefficient of the square structure fabric, taking into account the sum of the numerator and denominator of the fraction of the main weaves, is proposed. The research of the structure parameters of the square structure fabric depending on the weave, air permeability and surface density of the fabric is carried out. It has been determined that the increase in the number of yarns in the square fabric report and air permeability decreases the values of the average yarn number and fabric density. And for air permeability there is invariability of values of such parameters as air permeability coefficient, percentage of uncovered area of fabric and relative density of fabric. The technique of analysing the parameters of square structure fabric for experimental fabric samples has been developed. Parameters of square plain weave fabrics such as relative density of fabric, percentage of uncovered area of fabric, air permeability coefficient, pressure degree and air permeability index have been determined. Increasing the yarn twist ratio in the metric system, increases the uncovered fabric area percentage, air permeability coefficient and air permeability of the square weave fabric, while decreasing the relative fabric density and pressure degree index. The production intensity of square weaving on the loom with weave 1/1 (short overlaps) is higher relative to weaves 1/2,1/3 and 1/4 (long overlaps). The surface density of the fabric, fabric filling ratio (FFR) and fabric cohesion coefficient (**FCR**) of square structure increases with increasing yarn twist. Comparative researches of standard and developed suit fabric on the basis of their structural characteristics and physical and mechanical properties are carried out. The optimum parameters of fabric production with the help of RCCE have been developed: tension of*

main yarns - 20 cN (per 1 thread); the value of scalp - 15 mm; position of the scalp relative to the breast - (+25) mm. At these values of parameters breakage of the main threads will not exceed 0.05 breaks per 1m. Besides, such a combination of technological parameters allows to better preserve the strength properties of the thread in the fabric. The optimal combination of fibre mixtures for suit fabric in the weft, such as bamboo 90% and wool 10%, is substantiated. The advantages of the innovative suit fabric in terms of surface density, breathability, abrasion resistance, tensile strength, elongation and moisture absorption are shown.
Keywords: fabric, yarns, tension, design, breathability, bamboo, wool, coefficient, weave, density, methodology, equations, structure.

GENERAL CHARACTERISATION OF WORK

In the world wide research works are carried out to improve the technique and technology of production of finished quality products, to create its scientific basis. In this direction, much attention is paid to the creation of effective technologies that improve the quality and competitiveness of textiles, the development of methods for optimising textile productivity, the creation of high-performance technical means and technologies at textile enterprises. At the same time, in the production of textile products of different composition, taking into account their physical and mechanical properties, the determination of their quality indicators is one of the important issues. In this connection, special attention is paid to the development of design according to the given parameters to improve the physical and mechanical properties of suit fabrics. In this connection, in the USA, Japan, China, South Korea, Germany, Switzerland, India, Turkey, Russia, Uzbekistan and other developed countries, special attention is paid to improving the physical and mechanical properties of textile products of different composition. and improving their quality. Therefore, it is important to solve the issues of yarn preparation, providing further improvement of consumer properties of textile products. At the same time, to provide consumers with quality fabrics, one of the most important issues is the creation of competitive products based on the design of fabrics with specified properties of different composition, and the technology of their production.

The object of the study is the structure of suit fabrics and their hygienic properties (breathability).

The subject of the study is the design methodology and structure parameters of costume fabrics.

The aim of the research is to design with specified properties and production technology of suit fabrics.

Objectives of the study.

To address the task at hand, the following is outlined:

- To analyse the properties of costume textile materials such as fibre, yarn and fabric;
- to design suit fabrics with specified properties by breathability;
- to develop an innovative suit fabric;
- to investigate the influence of yarn and fabric parameters on the properties of innovative suit fabric;

- optimise the weaving process in the production of innovative suit fabrics.

Research methodology. To solve the set tasks in the work a complex method of research was used. In the theoretical part, analytical geometry methods were used in the study of tissue structure. In the experimental part of the work the methods of mathematical planning and analysis of results were used. Processing of the experimental results was carried out using the method of mathematical statistics. In the technological part the weaving machine "SOMET" was used. Standard devices for determining the hygienic properties of fabrics in the certification centre at TITLP "CentexUz" were used.

The scientific novelty of the study consists in the following:

- raw material composition for the production of suit fabric was substantiated;
- The method of calculation of a suit fabric with the given properties of air permeability is offered;
- variants of costume fabrics of the main weaves with constant rapport and number of yarn transitions in warp and weft are developed and investigated;
- parameters of suit fabrics structure were determined, innovative suit fabric was designed and developed according to the given parameters breathability;
- optimal technological parameters of suit fabric production on the basis of rotatable central compositional planning of the experiment of the second order have been obtained.

Practical results of the research consist in the following: samples of fabrics which have a good appearance, good air permeability, possess a combination of aesthetic-hygienic and physical-mechanical properties have been developed, which can be recommended to use these fabrics for clothes. Designing of fabrics with specified breathability properties on the basis of wool-bamboo yarn mixture allows to promptly introduce the results of work in the industry and solve the problem of suit fabrics production, which are in great demand in the Republic. The development of innovative costume fabric optimisation of technological parameters of production of these fabrics allows to reduce the breakage of threads on the basis and to improve the quality of produced fabrics. The methodology of designing fabrics with given properties, as well as analytical and experimental studies of the work can be used in the educational process when studying the course of structure and design of fabrics, methods and means of research of technological processes, as well as in the basics of theoretical and scientific research.

The validity of the research results is confirmed by empirical, mathematical models of technique and technology of suit fabrics production, consistency of the results of theoretical studies on the known evaluation criteria in the considered subject area with the data of experimental studies. The processing of the experimental results was carried out with the use of computers. The errors of direct and indirect measurements were calculated using the methods of mathematical statistics, the assessment of significance of the obtained results was determined with a confidence level of 0.95.

Scientific and practical significance of the research results. The scientific significance is characterised by the development of a methodology for designing suit fabric with a given breathability. It allows to make promptly dressing and production of suit fabrics with minimum expenses. Practical significance of the conducted research consists in the optimisation of the process, which allows to reduce thread breakage and improve the quality of fabrics.

Content of work

The introduction substantiates the relevance of the topic, formulates the purpose, objectives of the study, shows the scientific novelty and practical significance of the work, methodology, research and approbation of the work.

The first chapter provides a review of literature sources devoted to the research of fabric structure and parameters influencing their properties. The analysis of literary sources shows that the works are mainly oriented on the research of fabric structure, thus designing of fabrics is carried out on a certain thickness, on fabric filling, strength, phase order of fabric structure, linear densities of threads. The structure of clothing fabrics and their design are insufficiently studied, in particular, the design of clothing fabrics by air permeability is of topical importance. In the production of suit fabrics the influence of weave factors on the fabric structure is insufficiently studied.

The second chapter is devoted to the study of the properties of textile materials, where the characteristics and properties of fibres, yarns and fabrics are given. The comfort of clothing is determined by its ability to provide conditions for optimal heat exchange between the body and the environment. The assessment criteria in this case are the characteristics of the state of the underclothing air layer. When considering air exchange in relation to clothing, there are two possible ways in which this process can take place. The first is through the exposed areas of the garment, such as the neck, fastener, sleeve, etc. The second is directly through the materials of the garment. The second pathway is directly through the materials of the garment. The intensity of the process in the first case is determined by the construction of the garment, in the second case by the breathability of the fabrics. The study of the influence of fibre geometry on air permeability in textile materials shows that, in these materials, fibre characteristics play a dominant role. That is, the main element of the structure for air permeability is the fibre. Filtration capacity of materials is conditioned in correlation with the change of one of five fibre characteristics - cross-sectional shape, linear density, surface roughness, tortuosity, staple length. Based on the analysis of the data obtained, it can be stated that filtration efficiency is improved by: the use of fibres: with non-circular cross-sectional shape; the use of twisted fibres compared to non-twisted fibres; the use of thinner fibres compared to thicker fibres. The extent to which the fibre composition of materials affects air permeability depends on their porosity. As porosity increases, the importance of fibre composition as a factor in air permeability decreases. The influence of the fibre composition of fabrics and knitted fabrics on their air permeability increases in cases of changes in the structure of materials leading to a reduction in inter-strand pores. In cases where the inter-strand porosity (the inverse of the volume filling) is high enough, as well as at increased smoothness of the yarns, the prevailing factors of air permeability, regardless of the fibre composition, are the characteristics of the structure of the materials. From the analysis of raw material characteristics it follows that it is expedient to use blends of represented fibres of wool and bamboo, cotton, linen and polyester for production of suit fabrics. The estimation of yarn strength from multicomponent blends is given. For this purpose, a yarn blend consisting of two components with different geometric and strength properties is considered, assuming that the fibres in the yarn are uniformly distributed. The fibres are arranged in helical lines with a constant pitch, the pitch of the helical line being independent of the current yarn radius, in an arbitrary cylindrical layer removed at an arbitrary distance from the yarn axis, the fibres of all components have the same strain. The deformation of each component through tension ***T is*** expressed by the formula

T_i / E_i F_i (i=1,2), where E_i F_i is the stiffness of the components. From the condition of equality

deformations should $\frac{T_1}{E_1F_1} = \frac{T_2}{E_2F_2} \varepsilon_0 cos\theta T_1$ (2.1)

where cos Θ is the average value of the cosine angle over the cross-section, we take 0.95

From the equilibrium condition we find the total tension of each component

T_i, $T_1+T_2 = P \cos\theta$ (2.2)

The strength of a yarn blend is determined by the value of the ultimate force ***P*** at which the yarn fibres break away. The highest force occurs in the stiffest component, and after its detachment, there is a redistribution of tension between the components, and if the more deformable component will carry the load together with the blocked friction force, the yarn strength increases. Let us apply this method of yarn strength determination to two types of blends consisting of two

components.

1 type of mixture consisting of polyester (50%) and wool (50%) Let us denote by $E_p F_p$, and Esh B_{sh}, respectively the stiffnesses of polyester and wool, E_p, and E_{sh} the Young's moduli in tension,

$$F_p = \sqrt{\bar{T}_p/\rho_p}, \quad F_ш = \sqrt{\bar{T}_ш/\rho_ш},$$

Linear density T_p and T_{sh} *of* the components p_p and p_{sh} their densities in calculations we assume

$E_p = 4 - 109Pa$, $E_{sh} = 3 - 109Pa$, $T_p = 1.2 - 10\text{-}6$ kg/m, $T_{sh} = 4.5 - 10\text{-}6$ kg/m, $p_p = 1.32 - 103$ kg/m^3, $p_{sh} = 1.38 - 103$ kg/m^3, Thus we have $E_P F_P$ = 363 cH, E $F_{шш}$ = 978 cH, from the calculations it follows that the greatest stiffness of the mixture is equal to E F_{mm} = 978cH. Formula (2) taking into account (1), let us present it in the form P = 0.5 * $P = 0.5\% \cos P_{Esh\ Esh} (1 + E_p\ E_{p\ /EshEsh})$

We assume $with_{0\ EshEsh} = P0$ (P_O - detachment force determined experimentally). From experiments we have P0 = 489cN. Then for the strength of the mixture we obtain P = 335.13cN.

2 type of mixture consisting of bamboo (90%) and wool (10%). For bamboo we take $E_ъ = 1.5\text{-}10^9$ Pa, % = 1.2 - 10-6 kg/m, $p_p = 1.5 - 103$ kg/m^3, EbFb = 232 cH. The strength will be P = P0 (0.9 + 0.1Eb $_{Fb/Eb}$ F_m) = 447.7 cH. From the calculations, it can be seen that for the first type of mix, the strength of the mix is much less than the breakaway forces P0 = 489cH. For the second type of mix, the strength is slightly different from the allowable strength, indicating the influence of bamboo. In our case, variations of different blends of wool and bamboo are considered. The study of the effect of fibre geometry on air permeability in textile materials shows that, fibre characteristics play a dominant role in these materials. That is, the main element of the structure for air permeability is the fibre. Filtration capacity of materials is conditioned in correlation with the change of one of five fibre characteristics - cross-sectional shape, linear density, surface roughness, tortuosity, staple length. Based on the analysis of the data obtained, it can be stated that filtration efficiency is improved by: the use of fibres: with a non-circular cross-sectional shape; the use of twisted fibres compared to non-twisted fibres; the use of thinner fibres compared to thicker fibres. The extent to which the fibre composition of materials affects air permeability depends on their porosity. As porosity increases, the importance of fibre composition as a factor in air permeability decreases. The influence of the fibre composition of fabrics and knitted fabrics on their air permeability increases in cases of changes in the structure of materials leading to a reduction in inter-strand pores. In cases where the inter-strand porosity (the inverse of the volume fill) is high enough, as well as when the smoothness of the yarns is increased, the predominant factors of air permeability, irrespective of fibre composition, are the characteristics of the structure of the materials. Table 2.1 shows some characteristics of staple fibres of wool and bamboo linen, cotton and polyester (lavsan). Wool is the hair of animals that can be processed into yarn (Table 2.1). Wool fibres repel dirt and are easy to clean. Heat resistance -

(ability to retain heat) is one of the best known and favourite properties of wool. Wool has the highest heat retention properties. This action is due to the composition of its fibres to bind heat and retain it between the fibres. There is no other fibre like it in nature. The highest hygroscopicity is 18-25%. It absorbs moisture from the environment, but unlike other fibres it slowly absorbs and releases moisture, remaining dry to the touch. It swells strongly in water. Moistened fibre in a stretched state can be fixed by drying, when re-moistened, the length of the fibre is restored again. Good light fastness. Good stretchability. Good elasticity - not wrinkle resistant.

Table 2.1.

Characteristics of staple fibres of wool and bamboo linen, cotton and polyester.

Name	Indicators				
	Polyester	Cotton	Wool	Len	Bamboo
Range of fibre metric numbers (millitex) to be processed	1200 (840) - 6000 (170)	4500(220)- 9500 (105)	4500(220)- 9500 (105)	900(1100) 2700 (380)	1500 (680) - 6000 (170)
Specific gravity, g/cm3	1,38	1,50-1,52	1,32	1,43-1,50	1,50-1,53
Equilibrium moisture content (in %) of fibre at ambient humidity: 65% 95%	0,4-0,5 0,5-0,7	7,0-8,0 24-27	15-17 38-40	13 13	12,8-13,4 27-33
Breaking length under conditioned conditions, *km*	32-40	22-44	8,4-13,1	55-80	15-23
Tensile strength, kg/mm^2	44,2-55,2	33,4-67,0	15,8-19,8	82,5-120,0	22,8-35,0
Relative strength, gs/tex	32-40	25-35	20-30	60-70	32-40
Relative strength, %: in wet condition	98,5-100	110-120	80-90	100,0	98,0 - 100,0
Breaking elongation, %: under conditioned conditions wet	40,0-60,0 40,5-61,5	10,0-12,0 11,0-13,0	25-40 30-60	2,0-2,5 2,5-2,9	40,0-60,0 40,0-60.0
Shrinkage after wet processing, %	2,7-12,3	-	1,5	-	2,6-6,0
Tensile modulus, kg/mm^2	-	500-550	340-360	3000-5000	300-650
Degree of elasticity (in %) in tension: by 4% by 10%	- -	- -	78 71	- -	34,5-43,0 28,8-33,0

Fibre mass elasticity (angle of elastic recovery after removal of compressive load), *deg*: after 3 min after 60 min	102 - 105 107 - 110	74-79 85-89	114-120 121-126	- -	68-70 83-85
Resistance to repeated bending, number of cycles at a stress of 10 *kg/mm*^{2}	21300 30000	40000-50000	-	-	7000 16000
Resistance to ultraviolet irradiation (loss of strength after 20 hours of irradiation), %	-	50 (after 940 hours of sun exposure)	50 (after 1120 hours of sun exposure)	50(after 1000 hours of sun exposure)	20,0-35,0
Temperature areas, oc softening-melting crystallisation	235-255 220-210	Destroys at 1600	Becomes brittle at 100°, decomposes at 130°, charred at 250° - 300°	Becomes brittle at 100°, decomposes at 130°, charred at 250° - 300°	It won't melt
Heat capacity, *cal/g - deg.*	0,320	-	-	-	0,320
Temperature resistance, %: at +1400 loss of strength change of elongation	- 32-44 50-100	- - -	19-28 40 44	19-28 40 44	10-15
Double refractive power	0,17-0,22	-	0,010-0,011	0,010-0,011	-

Bamboo is an herbaceous plant of hot climates, the main feature of which is its high growth rate and unpretentiousness. Unlike cotton, it does not deplete the soil and does not require chemical treatment during cultivation. In addition, this high-growing grass has many useful properties, and its composition contains many valuable substances that contribute to health. The mechanical processing of bamboo consists of shredding and enzyme treatment, resulting in fibres up to 15 cm long. These somewhat rough fibres are called "bamboo linen", they are environmentally friendly and the healthiest fibres available. The traditional viscose production process involves alkali or carbon sulphur treatment, which is quick and fairly cheap. The material obtained in this way is called "bamboo viscose" or "bamboo rayon"; it is most commonly found, especially in blends with cotton. Bamboo fibre is a fluffy thread with a large number of cavities. It is very strong, has very good thermal insulating properties, is air permeable and can absorb a large amount of water, and when dyed, the colours are bright and saturated (Table 2.2). Importantly, bamboo fibre retains its ecological properties during disposal: discarded bamboo fibre decomposes like plant residues without polluting the environment. Table 2.2 shows the effects of environmental humidity on the equilibrium moisture content of yarns made of wool, bamboo, linen, cotton and polyester fibres.

Table 2.2.

Effect of ambient humidity on the equilibrium moisture content of yarns made from wool, bamboo, linen, cotton and polyester fibres.

№	Name	Equilibrium humidity, %				
		Polyester	Cotton	Wool	Len	Bamboo
1.	Ambient humidity at 65%.	0,5	8	17	13	13,4
2.	95% ambient humidity.	0,7	27	40	13	33

In Fig. 2.1, the abscissa axis corresponds to the ambient humidity in per cent (%) and the ordinate axis to the equilibrium moisture content of the yarn in per cent (%). The analysis of Fig. 2.1 shows that the equilibrium moisture content of polyester fibre is 0.5% at 65% ambient humidity and 0.7% at 95% ambient humidity. For cotton fibre, the equilibrium moisture content is 8% at 65% humidity, 27% at 95% humidity. The equilibrium moisture content is 13% for linen fibre at 65% and 95% ambient humidity. The equilibrium moisture content is 13.4% for bamboo fibre at 65% ambient humidity and the equilibrium moisture content is 33% at 95% ambient humidity. The equilibrium moisture content is 17% for wool at 65% ambient humidity and is an equilibrium moisture content of 40% at 95% ambient humidity.

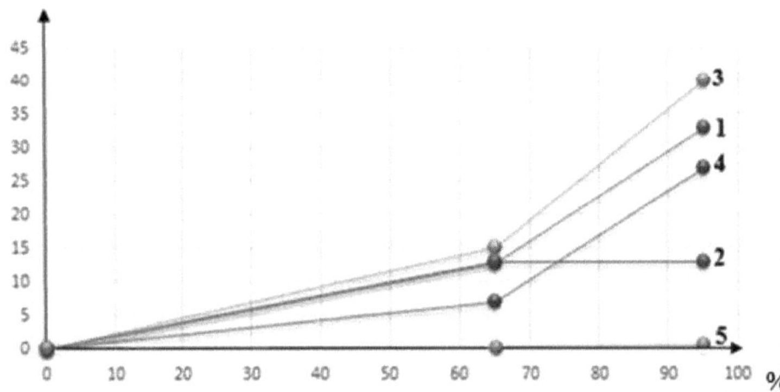

Figure 2.1. Graph of the effect of ambient humidity (%) on the equilibrium

moisture content (%) in yarn where: 1- bamboo; 2- linen; 3- wool; 4- cotton; 5- polyester.

In Fig. 2.2, the abscissa axis corresponds to the elongation of the yarn at break in per cent (%) and the ordinate axis to the relative breaking strength of the yarn in gs/tex.

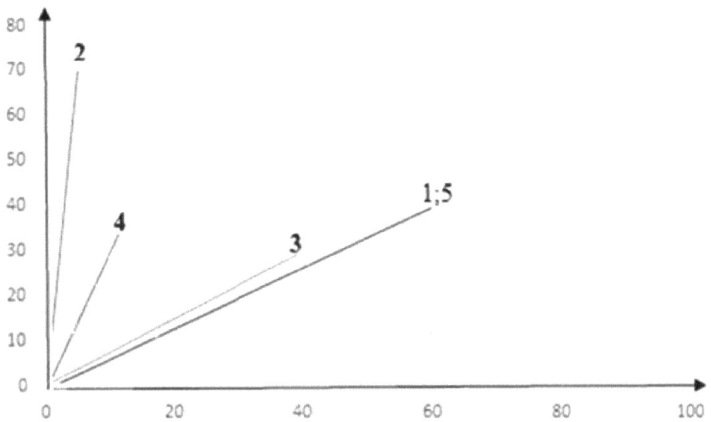

Figure 2.2. Graph of the effect of yarn break elongation on the relative breaking strength of yarn where: 1-bamboo; 2-linen; 3-wool; 4-cotton; 5-polyester.

The analysis of Fig. 2.2 shows that the elongation of break at 60% for polyester and bamboo has a relative yarn strength of 40 gs/tex. For wool at 40% elongation of break, the relative strength is 30 gs/tex. For cotton at 12% break elongation, the relative strength is 35 gs/tex. For linen yarn at 2.5 per cent break elongation, the relative strength is 70 gs/tex. On the basis of analysis of raw material properties it is expedient to use mixtures of woolen and bamboo, cotton, linen and polyester fibres in production of suit fabrics. The paper considers various variants of fabrics from blended wool and bamboo.

Table 2.3 shows the physical and mechanical properties of yarns such as artificial bamboo, wool, viscose, cotton and polyester.

Table 2.3.

Physico-mechanical performance of yarns.

Name of indicators	Artificial bamboo	Wool	Viscose	Cotton	Polyester
Relative strength, cH/tex	40-42	11-13	22-26	20-24	55-60
Elongation, %	14-16	25-30	20-25	7-9	25-30

Relative wet strength, cH/tex	34-38	8-10	10-15	26-30	54-58
Elongation in wet condition, %	16-18	27-35	25-30	12-14	25-30

The analysis of Table 2.3 shows that in dry and wet condition, the relative strength of bamboo yarn is four times that of woollen yarn, while being inferior only to polyester yarn.

Table 2.4 shows the physical and mechanical properties of bamboo-wool blended yarn in different combinations.

Table 2.4.

Physico-mechanical parameters of bamboo-wool blended yarn.

№	Percentage of bamboo investment in wool for yarns	Breaking strength, N	Elongation, %	Moisture absorption, %
1	Bamboo:wool-90:10%	260	24	77
2	Bamboo:wool-70:30%	220	21	74
3	Bamboo:wool-50:50%	200	20	70
4	Bamboo:wool-30:70%	190	19	68
5	Bamboo:wool-10:90%	170	17	67

The analysis of Table 2.4 shows that with the increase in the percentage of bamboo inclusion in wool, the physical and mechanical parameters of blended yarn bamboo - wool increase breaking strength by 35%, elongation by 3%, moisture absorption by 13%.

In blended fabrics with a predominance of bamboo should be noted: high environmental friendliness of the raw material; excellent hygienic and health qualities; hypoallergenic; high durability, and at the same time softness, elasticity and breathability; environmentally friendly bamboo fibre kills up to 70% of bacteria contacting it in a day, and this effect lasts up to 5 washings; clothing containing bamboo retains up to 100% of ultraviolet radiation; clothing is very pleasant to the touch, never causes scuffs and irritations and promotes their healing; has an increased ability to absorb water and unpleasant odours; practically does not wrinkle, is well washed and retains an attractive appearance and consumer properties up to 500 washings. Thus, in such clothes it is impossible to sweat and overheat, it will reliably protect from cold and sun, get rid of allergic and other irritations and at the same time it is durable and comfortable enough to wear. A special selection of suit fabrics contributes to the creation of such a reliable suit image. Of course, we are not talking about any impact-resistant properties of these fabrics, but the state standard for suit fabrics contains an extensive list of indicators that they must meet. Suit fabrics do not have a strictly defined composition and consist of natural and chemical fibres. The requirement of safety of synthetic raw materials, dyes, other chemical substances used for manufacturing of men's suit fabrics is mandatory. Without regulating the composition of suit fabrics, the state standard defines mandatory requirements for their properties. One of the most important parameters determining the operational properties of the fabric is durability, which provides reliability of the fabric during the whole service life of the suit. Taking into account the conservativeness of fashion, the service life of a suit is from one to three years. The strength of the fabric is characterised by its breaking load. A 5cm wide and 20cm long strip of fabric should be able to withstand a tensile load of at least 34.3kG. During wear the suit is exposed to various factors, including weather conditions, and washing is not an exception, so the next requirement is the consistency of shape. It is achieved by a low permissible value of change of the suit fabric dimensions after wet processing, namely, only minus 2 per cent on the fabric base (in the grain direction) and plus 1.5 per cent on the weft (in the cross direction). In a number of indicators of operational properties it is necessary to note lightfastness of fabrics, i.e. resistance of fabrics colouring to the action of light. For suit fabrics of especially durable colouring such indicators as the degree of colouring resistance to sweat, washing, friction, ironing and organic solvents are at the level of four-five points on a five-point scale. It is important that during the entire service life of the suit, in addition to high performance characteristics, the fabric also retains its aesthetic properties. One of such characteristics is the non wrinkling and pilling of fabrics, i.e. the tendency to form lint on the fabric, which gives it a sloppy look. This indicator for suit fabrics is not more than four pills per $10cm^2$. Suit refers to those types of clothing in which a person is during the whole working day, and often during the rest, that is about 8-10 hours daily, and all this time he should feel comfortable. Ergonomic properties of fabrics determine the index of breathability of fabrics. This property ensures the exchange of air under the suit with the ambient air, in other words, ventilation of the underclothing air. The air permeability index of suit fabrics is not less than 50 dm /m^{32} sec. This means that $1m^2$ of suit fabric should permeate a volume of air of at least 50 dm^3 per second. Whether this is a lot or a little. For the entire variety of textile materials, this figure ranges from 3 to 2000 dm /m^{32} approximately. Since one of the many functions of a suit is to keep warm, a lower limit of at least 50 dm /m^{32} , is sufficient to keep warm and feel comfortable. The upper limit is not limited. It is clear that for summer suits made of lightweight fabrics the air exchange values can be higher. Even a brief review of the regulating indices of suit fabrics allows us to conclude that their totality can be achieved only in blended fabrics. We have to admit it, no matter how much we would like to have clothes only from natural fibres - linen, cotton, wool. Fabrics made of linen, cotton, of course, are very comfortable, but give shrinkage, easily crumpled. Although today the trend of "slight wrinkling" is still in fashion, it is unacceptable for a business suit, as it destroys the image of a successful dynamic man. Fabrics made of pure wool and semi-woolen fabrics have long been used for suits. They include worsted fabrics with a clear pattern of interlacing threads:

- boston is a pure wool fabric, smoothly dyed;
- gabardine is a pure woollen fabric with a diagonal weave;
- Cheviot is a semi-woolen fabric (cotton on the base, wool on the weft);
- crepe - pure woollen or semi woollen fabric of crepe weave;
- Tricot - pure woollen or semi woollen fabric of patterned weave.

The listed fabrics are in demand in the market nowadays, but the main segment of the suit fabrics market is made up of blended fabrics, where chemical fibres are present, at least in the base of the fabric. This allows to make the fabric stronger, reduce shrinkage, increase lightfastness and, importantly, reduce the price. Modern technologies allow to produce fabrics based on mixed fibres indistinguishable in appearance from fabrics made of natural raw materials. Only the label on the suit objectively reveals to us all the components of the suit fabric. Wool or cotton, specified in the composition, involuntarily cause satisfaction, we know that it is good, it is comfortable. Then, as a rule, follows the abbreviated name of the chemical fibre that adds a number of useful, practical properties to it:

- the addition of up to 50% polyester (PE) and polyacrylonitrile (PAN) fibres increases the shape stability of fabrics,
- The addition of up to 40% polyester fibres (PE) reduces the tendency of fabrics to form pills (lint),
- The addition of up to 4% of capron and lavsan increases wear resistance.

Fabric made from bamboo fibre is light, soft and has a pleasant natural lustre - superior to natural silk. The fabric has a high elasticity, which makes it virtually wrinkle-free, and a high resistance to wear: the tensile strength of bamboo fibre is comparable to steel. Bamboo fabric does not cause allergic reactions, does not irritate the skin, protects it from ultraviolet light (reflecting 98% of harmful rays), has antibacterial properties and prevents the reproduction of pathogens, fungi and dust mites (bamboo fibre kills 70% of bacteria), and retains these antibacterial properties even after a hundred washings. The amino acids contained in bamboo have a favourable effect on the energy balance of the skin, and its fabric has an anti-inflammatory effect on the body. Bamboo fibre is non-electrifying, has excellent thermo-regulating properties, transmits 20% more air and absorbs 60% more moisture than cotton fabrics, resulting in high hygienic properties. Bamboo fabric is easy to dye and retains its colour perfectly. The eco-friendly method of producing bamboo yarns is similar to that used for linen and hemp: the bamboo stalks are crushed and natural enzymes are used to crush them, allowing the fibres to separate. In ancient times, bamboo was used to make paper in the same way. In the industrial process, the fibres are separated chemically - using alkalis, carbon disulphide and acids - and then extruded using mechanical devices. For the past few years, bamboo has been the main fibre material of the textile industry as a whole and is popular with leading designers. Bamboo fabrics and bamboo/cotton blends are used to make bed linen, dressing gowns, evening and casual dresses for women, and to knit light jumpers and socks. Coats and jackets are sewn from wool-blend fabrics, warm knitted clothes are made. Consequently, in blended fabrics with predominance of bamboo should be noted: high environmental friendliness of the raw material; excellent hygienic and health-improving qualities; hypoallergenic; high strength, and at the same time softness, elasticity and breathability; environmentally friendly bamboo fibre kills up to 70% of bacteria contacting it in a day, and this effect is maintained up to 5 washings; clothing containing bamboo retains up to 100% of ultraviolet radiation; clothing is very pleasant to the touch, never causes scuffs and irritations and promotes their healing; has an increased ability to absorb water and unpleasant odours; practically does not wrinkle, is well washed and retains an attractive appearance and consumer properties up to 500 washings. Thus, in such clothes it is impossible to sweat and overheat, it will reliably protect from cold and sun, get rid of allergic and other irritations and at the same time it is durable and comfortable enough to wear. What is important, this clothing retains its ecological properties and in the process of disposal: the discarded thing from it decomposes as well as plant residues, without polluting the environment. To summarise, it should be noted the properties of fabrics for suits should have comfort and practicality.

For the study of blended fabric as a base we took fabric art. 23195, where blended yarns 50 % wool and 50 % roluester are used in the base, and blended yarns wool - bamboo with different ratios are used in the weft.
The results of the fabric variants are summarised in Table 2.5.

Table 2.5.

Results of fabric variants.

№	Fabrics	Yarn blend composition,%, weft	Moisture absorption, %	Breaking strength, N	Air permeability, cm /cm^{32} sec	Abrasion resistance, cycles
1	Basic version	50% Wool 50% Polyester	40	300 245	11	4060
2	1 fabric option	90% Wool 10 Bamboo	67	240 207	19	4010
3	2 variant	70% Wool 30 Bamboo	68	250 210	28	4030
4	3 option	50% Wool 50 Bamboo	70	270 200	30	4050
5	4 option	30% Wool 70 Bamboo	74	280 220	39	4080
6	5 option	10% Wool 90 Bamboo	77	300 260	48	4100

where: numerator by warp; denominator by weft.

In the third chapter the subject of research is one of the properties of textile materials, providing their comfort - air permeability. This property, moreover, for materials and products of a certain purpose, in particular clothing fabrics, can be the main one determining their quality. Up to now the research of air permeability of fabrics has been directed to the study of the dependence of the velocity of liquid or gas passing through a porous material, which show that the character of the flow velocity dependence on the head is no longer linear and it is reflected in the equation called "Rakhmatullin's equation", where: the coefficient of air permeability used at a certain value of pressure drop is proposed; the classification of textile materials according to their air permeability is developed; the degree function is obtained. The influence of various factors on air permeability has also been studied, in particular the influence on air permeability of material density, the nature of fibre distribution in the material, the type of weave of the fabric, the twist of threads, the surface density of woven materials, the filling of woven materials with fibre, material density and fibre diameter, and the geometrical characteristics of fibres. The last two studies then had a practical implementation, which consisted in the creation of the Micronair fibre fineness evaluation device, which in a modified form is part of the integrated Spinlab for the evaluation of fibre properties. Thus, the dependences of air permeability of fabrics on these or those characteristics of their structure have been investigated, therefore at the present stage it is necessary to develop calculations of air permeability and forecasting of this property for newly designed garment fabrics. In addition, air permeability is one of the main parameters, which allows a first approximation of the stress of fabric production on the weaving machine. The air permeability is influenced by the type of raw material (yarns), the linear yarn density, the shape of its cross-section, the weave of the fabric, the warp and weft densities, the order of the fabric construction phase, the threading and weaving parameters on the weaving machine. Therefore, the criterion of fabric structure assessment is air permeability, which determines the quantitative and qualitative assessment of fabric structure. In the work on the basis of the given air permeability of the fabric the method of its designing is developed. The coefficient of air permeability, the percentage of unfilled and filled with fibrous material area of the fabric, the average yarn number (linear density), the fabric density and the equation of air permeability for a given fabric are determined. Fabric filling parameters were developed and prototypes of non-square fabric were developed. Comparative researches of designed and experimental fabrics on air permeability are carried out. At designing of garment fabrics with the set properties on air permeability, air permeability *B is* used as a criterion of estimation of the structure of garment fabrics. Therefore the method of its designing is offered on the basis of the set air permeability *B of* fabric. The following parameters of the square structure fabric of the main weaves are determined, such as the air permeability coefficient ***C***, the percentage of unfilled and filled area of the fabric with fibrous material *f,* the average yarn number *N, the* fabric density *P.* The equation of the air permeability coefficient of the square structure fabric, taking into account the sum of the numerator and denominator of the fractions of the main weaves, is proposed. The research of the structure parameters of the square structure fabric depending on the weave, air permeability and surface density of the fabric is carried out. It is determined that the increase in the number of yarns in the square weave and air permeability decreases the values of the average yarn number and fabric density. And for air permeability we have invariance in the values of parameters like air permeability coefficient, percentage of uncovered area of fabric and relative density of fabric. Air permeability can be calculated with sufficient accuracy using the equation:

$$B = Ch^{\tau} \quad (3.1)$$

where *C*- air permeability coefficient; ***h*** - rarefaction behind the fabric in mm of water. st, in our case we take ***h* =5** mm of water. st; τ - index of the degree of air pressure.

The task is to determine the air permeability coefficient ***C*** from the given air permeability ***B*** to find the proportion of uncovered area ***f*** % to calculate the relative fabric density ***E*** and the yarn number ***N***. From the formula (3.1) we find the air permeability coefficient

$$C = \frac{B}{h^{\tau}} \quad (3.2)$$

We determine the percentage of uncovered area of the square structure fabric by the formula

$$C = 0.00588 \cdot f^{2.46} \qquad (3.3)$$

Moreoverf=fo -fy, we find $f_o = f_y = \sqrt{f} \quad (3.4)$

Relative densities

$$E = 100 - f_o \; (3.5)$$

Determine the average yarn number.

$$N = \left(\frac{0{,}2 \cdot E \cdot c \cdot \mu}{M_T}\right)^2 \qquad (3.6)$$

where: ***E*** - relative density of fabric in %; **c** - coefficient of maximum density of fabric, equal for yarn 80; μ - coefficient of attraction in surface density of fabric equal to 1,05; *Mt* - surface density of 1 m^2 of fabric in gr.

Determining the density of a square structure fabric

$$P_{y=} \frac{M_T \cdot N}{21} \qquad (3.7)$$

Since *N* = -t^, after substitution into formula (3.7) we have

$$P_{y=} \frac{M_T \cdot 1000}{21 \cdot T} \qquad (3.8)$$

T - linear yarn density defined in texas.

The density of fabrics determines their structure and many properties. Thus, the denser the fabric, the stronger the pressure of warp and weft yarns on each other, the greater the friction forces between their constituent fibres, and therefore, the stronger the fabric to tear and higher its resistance to abrasion. With increasing density increases the thickness, weight and stiffness of the fabric, reduces its stretchability, drapeability, shrinkage when soaking and washing. As the density decreases, the porosity of the fabric increases, thus increasing its air permeability, moisture absorption capacity and thermal protection properties, but decreasing its wind protection properties. The equation of air permeability for a given fabric is as follows

$$B = M(\sqrt{h+K} - \sqrt{K}) \quad (3.8)$$

$$M = \frac{C+50}{1{,}088} \quad (3.9)$$

$$K = \frac{1080}{C^2} \quad (3.10)$$

The design fabric of square structure on structure and air permeability should correspond to the experimental sample. We proposed to develop a fabric of square structure of surface density M_T = 240 g/m^2 with air permeability B = 50cm /cm^{32} sec.

1.Determine the air permeability in the adopted values B = 50cm /cm^{32} sec.

2.Find the air permeability coefficient C from equation (3.2), C = 10.7.

3.Determine the percentage of uncovered area of the fabric from the formula (3.3) From where we get f = 0.124. Find by (4) $f_o = f_y = \sqrt{0{,}124} = 0{,}35 \cdot 100 = 35.$

Relative densities E = 100 - 35 = 65 %.

Hence, the uncovered area of the fabric is 35%, then the relative density of the fabric is E = 65%.

4.Determine the yarn number according to formula (3.5)

$$N = \left(\frac{0{,}2 \cdot 65 \cdot 80 \cdot 1{,}05}{240}\right)^2 = 20$$

Determines the fabric density by formula (3.6)

$$P_y = \frac{240 \cdot 20}{21} = 220$$

yarn/dm

The equation of air permeability for the given fabric is determined by formulae (3.7) (3.8) and (3.9).

$$B = 55{,}8\left(\sqrt{5+9{,}4} - \sqrt{9{,}4}\right) = 40{,}2 \quad M = \frac{10{,}7+50}{1{,}088} = 55{,}8 \quad K = \frac{1080}{10{,}7^2} = 9{,}4$$

The designed fabric corresponds to the experimental sample in terms of structure and breathability, which are shown in Table 3.1.

Table 3.1.

Filling parameters of the fabric

№	Name	Unit.	Indicators
1	Finished fabric width	see.	142

2	Filling width across the reed	see.	152
3	Main yarn number (linear density)	(tex)	20 (50)
4	Weft yarn number (linear density)	(tex)	20 (50)
5	Fabric density on the base	n/dm	320
6	Weft density	n/dm	150
7	Number of warp threads to be inserted in the reed tooth	pcs.	4
8	Reed number	tooth/dm	80
9	Surface density of the fabric	gr./m^2	240
10	Fabric filling	%	65
11	Air permeability of the fabric	cm /cm^{32} s	48

In the training and testing laboratory "CENTEXUZ" at TITLP on the devices "AP-360SM" the air permeability of the experimental sample of fabric was determined. Table 3.2 shows a brief technical characteristic of the device.

Table 3.2.

Brief technical characteristics of the device

Device name	AP- 360SM air permeability testing device.
Appointment of the device	To determine the degree of breathability of different types of fabrics.
Working conditions on the device	Room temperature-20±2^0 C. Relative humidity 65±2%.
Test range	0.5-390 cm /cm^{32} -sec.
Area of the vacuum chamber	38.3cm.^{2}
Range of inclined pressure gauge	0.1-300 mmHg (hydrostatic pressure).
Vertical pressure gauge range	0.1-400 mmHg (hydrostatic pressure).
Sample requirements	The dimensions of the sample are 160x160 mm.
Special requirements	Depending on the thickness of the tissue, the required diaphragm is selected.
Notes	Before using the appliance, the water level "0" must be checked.

We used interchangeable calibrated diaphragms with aperture diameter of 16 mm. In the final operation we determined the air permeability index of the tested fabric sample using a special table. Such measurements and calculations were made at least five times. Determination of the numerical characteristics of the fabric air permeability was carried out according to the known method in the following sequence, presented below.

1 .Average air permeability of the fabric $\bar{Y}=\frac{1}{m}\sum_{i=1}^{m}Y_i$

2 .Dispersion $S^2\{Y\} = \frac{1}{m-1}\sum_{i-1}(Y_i - \bar{Y})^2$

3 .Mean square deviation $S\{Y\} = \sqrt{s^2\{Y\}}$

4 .coefficient of variation $C\{Y\} = \frac{S\{Y\}}{\bar{Y}}100$

5 .Absolute confidence error of the mean value $E\{\bar{Y}\} = S\{\bar{Y}\}\frac{t_T}{\sqrt{m}}$

where: t_T *{PD = 0.95, /= m-1=5-1=4} = 2.776quantile of* Student's distribution.

6 .Relative confidence error of the mean value

$$\delta\{\bar{Y}\} = C\{Y\}\frac{t_T}{\sqrt{m}}$$

where: t_T *{P_D = 0,95, f= m-1=5-1=4} = 2,776* quantile of Student's distribution.

The error of the obtained values was within 5%. The results of calculations are given in Table 3.3

Table 3.3.

Numerical characteristics of fabric breathability

Name	Fabric breathability values					
	Average value of Y	Дис_-Persia S $\{Y\}^2$	Mean square deviation S{Y}	Coefficient of variation C{Y}	Absolute error of the mean E {Y}	Relative error of mean value δ{Y}
Experimental fabric	48	2,5	1,6	3,3	2,0	4,0

For square fabrics, the warp and weft yarn numbers and the warp and weft densities have the same values. Consequently, the warp and weft linear densities are the same. Without disturbing this balance it is possible to obtain, with unchanged yarn number of warp and weft and variable yarn density of warp and weft, a similar fabric of non-square structure. And if one yarn system is increased by ***n*** times, the other yarn system must be reduced by ***n*** times. For the similar non-square fabric we have at n = 1,45; P_o = 150 n/dm; R_u = 320 n/dm;

N_o = ***Ny*** = 20 or $T_o = T_u$ = 50 tex. The difference between the experimental values of air permeability and the calculated values of the equation for this fabric is 16%. This is due to the weave of the fabric, composition, type and twist of yarn. The deviation of air permeability of the fabric is 2 cm /cm^{32} sec, or 4%, which is quite acceptable. The deviation occurred due to the construction of fabric of non-square structure, i.e. increase in the value of fabric density on the base and decrease in the density of fabric on the weft. To determine the air permeability of garment fabrics of the main weaves, the following parameters of the fabric of square structure of the main weaves have been determined, such as the air permeability coefficient, the percentage of unfilled and filled area of the fabric with fibrous material, the average yarn number, the fabric density. The equation of air permeability coefficient of square weave fabric is proposed, taking into account the sum of numerator and denominator fractions of the main weaves. The research of the structure parameters of the square structure fabric depending on the weave, air permeability and surface density of the fabric is carried out. It is determined that the increase in the number of yarns in the square weave and air permeability decreases the values of the average yarn number and fabric density. And for air permeability we have unchanged values of such parameters as air permeability coefficient, percentage of uncovered fabric area and relative fabric density. Depending on the structure of the fabric surface, fabrics are divided into smooth, pile and felted. Smooth fabrics are those that have a clear pattern of the main weaves, such as plain, twill and satin. In the process of finishing, smooth fabrics are usually scorched on the front side. Depending on the type of weave, density, and degree of warp and weft curvature, the surface of the fabric may be dominated by warp or weft yarns. Equally supported fabrics have the same area of warp and weft overlaps on the front side (Fig. 3.1). In weft-supported fabrics, the weft overlaps predominate on the face (Fig. 3.2), while in warp-supported fabrics the main overlaps predominate (Fig. 3.3). In smooth fabrics, the support surface is formed by the projecting ridges of the yarn waves. The weave has a significant effect on the area of the support surface of the fabric: the longer the overlaps, the larger the area of the support surface. When a fabric abrades, its support surface is the first to be destroyed. Fabrics with a larger surface area are slower to deteriorate from abrasion. In addition, the support surface area affects the breathability of the fabric.

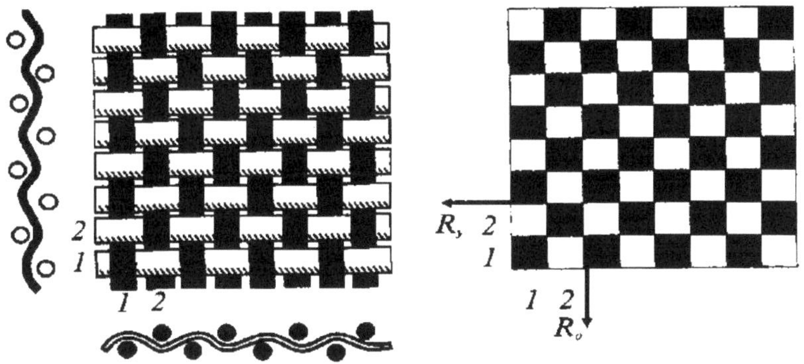

Fig.3.1. plain weave, overlapping 1/1.

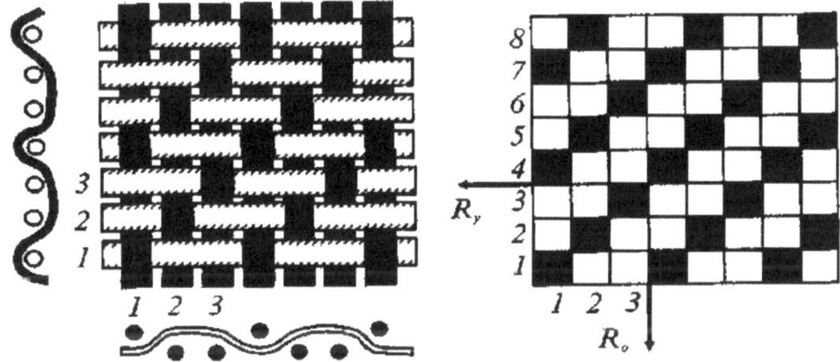

Fig.3.2 Twill weave, 1/2 overlap.

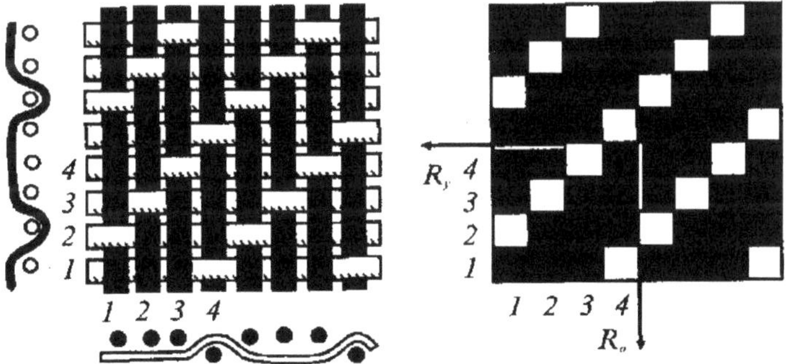

Fig.3.3. Twill weave, overlap 3/1.

Air permeability is an important characteristic of fabrics, expressed as the ability to allow air to pass through and ensure that the garment will be well ventilated, maintaining a certain ratio of humidity and gas composition of the air layer under the material. Carbon dioxide tends to accumulate in the space under the garment. Its concentration is twice as high as in normal air. If the content of this substance in the space under the clothes is 0.1 per cent, fainting may occur. A person begins to tire quickly and feel very tired. That is why it is so important that the fabric is well ventilated and its structure is porous. A common characteristic of breathability is the air permeability coefficient. The air permeability coefficient of a material shows the amount of air passing through 1 m^2 of fabric in 1 sec at a certain pressure difference on both sides of the material. The work was carried out in the following sequence. We determine the air permeability coefficient of plain weave fabric for fabrics of square structure.

$$C_n = \frac{B}{h^{\tau}} \quad (3.11)$$

The percentage of uncovered fabric area is determined from the formula

$$Cn=0{,}00588 \cdot f2.46 \quad (3.12)$$

Determine the average yarn count for square fabrics

$$N = \left(\frac{0{,}2 \cdot E \cdot c \cdot \mu}{M_T}\right)^2 \quad (3.13)$$

where: ***E*** - relative density of fabric in %; **c** - coefficient of maximum density of fabric, equal for yarn 80; μ - coefficient of attraction in surface density of fabric equal to 1,05; M_T - surface density of 1 m^2 of fabric in gr.

Determine the density of a square structure fabric

$$P_{y} = \frac{M_T \cdot N}{21} \quad (3.14)$$

Equation of air permeability coefficient of *cc* twill weave fabrics for square weave fabrics

$$C_C = \frac{0{,}263 \cdot \sqrt{n} \cdot C_n}{f_g \cdot N^{0{,}0625}} \quad (3.15)$$

where *Cn* - air permeability coefficient of plain weave fabric for square weave fabrics; ***n*** - number of yarns in the raport of square twill fabric; f_g - live cross-section of square weave fabric (it is within 0.2-0.4); ***N*** - average yarn number in square weave fabric.

Let it be proposed to produce a fabric of square structure with air permeability ***B*** = 50cm /cm^{32} sec. According to the formulas (3.11) - (3.15) we will carry out the calculation, the results of which are presented in Table 3.4.

Table 3.4 shows the results of calculating the square fabric parameters for the main overlapping weaves 1/1,1/2,1/3,1/4,1/5,1/6 and 1/7 (plain and twill).

Table 3.4.

Results of calculation of the square structure fabric parameters for the main weaves.

№	Name	The number of threads in the raport of a square structure fabric.						
		2	3	4	5	6	7	8
1	Air permeability coefficient, ***C***, cm /cm^{32} sec.	10,7	13,4	15,5	17,4	19,0	20,5	21,9
2	Percentage of uncovered fabric area, ***f***, %	35,2	36,7	37,8	38,6	38,6	39,8	40
3	Relative tissue density ***E***, *%*	64,8	63,3	62,2	61,4	61,4	60,2	60
4	Average yarn number, ***N***	20,6	19,6	19,0	18,5	18,5	17,8	17,6
5	Fabric density, ***P***, yarn/dm.	235	224	217	211	211	203	201

At constant value of surface density of square fabric ***MT***= 200gr/m^2 and variable value of air permeability of fabric ***B*** cm /cm^{32} sec the parameters of square fabric were determined. The calculation results are presented in Table 3.5.

Table 3.5.

Results of calculation of square fabric parameters at ***MT***= 200gr/m^2

№	Name	Air permeability ***B***, cm /cm^{32} sec.				
		50	100	150	200	250
1	Air permeability coefficient, ***C***, cm /cm^{32} sec.	10,7	21,3	32,0	42,6	53,3
2	Percentage of uncovered fabric area , ***f***, %	35	40	43	45	45
3	Relative tissue density ***E***, *%*	65	60	57	55	55
4	Average yarn number, ***N***	29,8	25,4	22,9	21,3	21,3
5	Fabric density, ***P***, yarn/dm.	283	242	218	203	203

At constant value of fabric breathability ***B*** = 50cm /cm^{32} sec^2 and variable value of surface density of square structure fabric ***MT*** gr/m, the parameters of square structure fabric were determined. The calculation results are presented in Table 3.6.

Table 3.6.

Results of calculation of square structure fabric parameters at air permeability of fabric ***B*** = ***50*** cm /cm^{32} sec

№	Name	Surface density of fabric ***Mt***, g/m^2				
		50	100	150	200	250
1	Air permeability coefficient ***C***, cm /cm^{32} sec.	10,7	10,7	10,7	10,7	10,7

2	Percentage of uncovered fabric area *f*, %	35	35	35	35	35
3	Relative density of the fabric *E, %*	65	65	65	65	65
4	Average yarn number, ***N***	477	119	53	29,8	19
5	Fabric density ***P,*** thread/dm.	1135	566	379	284	226

The analysis of Table 3.4 shows that an increase in the number of yarns in the square structure fabric report leads to an increase in the air permeability coefficient of the fabric and the percentage of uncovered area of the fabric**,** and to a decrease in the relative density of the fabric**, *the*** average yarn number and the density of the fabric. Table 3.5 shows that as the air permeability of the square structure fabric increases: firstly, the air permeability coefficient and the percentage of uncovered fabric area increase: secondly, the relative fabric density, average yarn number and fabric density decrease. Table 3.6 shows that an increase in surface density of the fabric causes unchanged values of air permeability coefficient, percentage of uncovered fabric area and relative fabric density in all variants, while the values of average yarn number and fabric density decrease.

To determine the influence of yarn twist on the air permeability of garment fabrics on the basis of an experimental fabric sample, the parameters of square structure fabrics are evaluated. The technique of analysis of the experimental sample of fabric is developed. Where the following parameters of fabric of square structure of plain weave are determined, such as relative density of fabric, percentage of uncovered area of fabric, air permeability coefficient, index of pressure degree and air permeability. Parameter values of square weave fabric were obtained as a function of yarn twist ratio in metric system. With the increase of yarn twist ratio in metric system, the percentage of uncovered area of fabric, air permeability coefficient and air permeability of square structure fabric increase, while the relative density of fabric and pressure degree index decrease. The air permeability of fabric is of great importance in the evaluation of hygienic and technical properties. At estimation of hygienic properties of linen fabrics the characteristic of their air permeability is obligatory, and for innovative fabrics the index of air permeability is introduced in technical conditions. When assessing clothing fabrics, the air permeability indicator characterises windproof properties of the fabric. For parachute fabrics air permeability is an indicator taken into account when calculating the design and operational properties of the parachute. For fabric filters air permeability of applied fabrics is one of the most important indicators. In spite of such great importance of air permeability, this property of fabric is insufficiently studied, the method of air permeability determination requires clarification. Dependences of air permeability on parameters of fabric structure are less studied. Establishing relationships between the structure of the fabric and the value of air permeability is one of the most difficult tasks, as the air permeability index is influenced by many factors - toning (linear density), the shape and volume weight of yarn, the density of the fabric on the base and on the weft, the weave of the fabric, the texture of the front side and back of the fabric, the type of finishing and others. The main parameter of fabric structure in terms of air permeability is the fabric density and in this connection the greater or lesser coverage of the fabric area by yarns, or the total area of the so-called through pores in the fabric sample. The size and twist direction of the yarn used is also of great importance. The twist has a significant influence on the yarn structure. When twisting, the fibres are arranged in helical lines of variable pitch and radius. Each fibre does not lie in a single yarn layer along its length, but in a number of layers, passing from the yarn centre to the periphery and back again. The fibre sections in the outer layers of the yarn are more tense than those in the centre of the yarn. This creates an unbalanced structure, as a result of which the yarn winding off the cob or bobbin twists and loops. The inclination of the coils lying in a particular yarn layer is subject to fluctuations and changes with the yarn diameter. Thus, the less uniform the yarn thickness, the more unevenly the twists are distributed along the length of the yarn. Depending on the requirements, the single yarns produced during spinning can have a weak or strong twist. A weak twist results in a less strong but softer yarn, while a strong twist results in a dense and stiff yarn. Depending on the direction of twist, the twist is labelled with the Latin letters Z and S. With Z twist, the turns run from bottom left to top right, with S twist from bottom right to top left. The twist direction of the yarn influences the appearance of the materials produced from it. When the warp and weft yarns are twisted in one direction, the yarn turns in the fabric are arranged in different directions, resulting in a more relief weave pattern. When warp and weft yarns are twisted in different directions, the fibres in the fabric are arranged in the same direction and the texture of the fabric has a plaid appearance. This causes a reduction of pores in the fabric, which are located between the warp and weft yarns. Therefore, it is reasonable to study the influence of yarn twist size and direction on fabric air permeability. The work was carried out in the following sequence. We determine the relative density of the fabric.

$$E = \frac{M_T\sqrt{N}}{0,2 \cdot c \cdot \mu} \qquad (3.16)$$

where: N - average metric number of yarn; **c** - coefficient of maximum fabric density, equal for yarn 80; μ - coefficient of attraction in surface density of fabric equal to 1,05; M_T - surface density of 1 m^2 of fabric in gr.

Then we determine the percentage of uncovered fabric area $f = f_0 - f_y$ (3.17)

Having determined the logarithm of the percentage of uncovered fabric area **lg** f, we calculate the logarithm of the air permeability coefficient by the equation

$$\lg C = \lg f - 2.46 + \lg 0.00588 \quad (3.18)$$

We then determine the air permeability coefficient C.

For air permeability coefficient C from 1 to 100, the degree τ of pressure is determined by the empirical formula

$$\tau = 0,5 \cdot \left(1 + \frac{1}{1 + 0,056 \cdot C}\right) \quad (3.19)$$

The air permeability of a square fabric is determined by the equation

$$B = Ch^{\tau} \quad (3.20)$$

where C ***is*** *the* air permeability coefficient; h ***is*** *the* rarefaction behind the fabric in mmHg; τ *is the* pressure degree index.

Produced fabric square structure plain weave, where the fabric density P = 220 threads / dm., the surface density of the fabric M_t = 240 g / m^2 , yarn number N = 20 (linear density T = 50 tex) on the warp and weft, and the weft used different twist (yarn twist coefficient in the metric system **an** varied from 50 to 100). According to formulas (3.16) - (3.20) we carry out calculations, the results of which are presented in Table 3.7 depending on the yarn twist ratio.

Table 3.7.

Results of calculation of the square structure fabric parameters depending on the yarn twist coefficient.

№	Name	Yarn twist ratio in metric system a_n, (in tex system a_m)					
		50 (16)	60 (19)	70 (22,1)	80 (25,3)	90 (28,4)	100 (31,6)
1	Average metric yarn number N (linear density T, tex) in the fabric	20 (50)	19,9 (50,2)	19,8 (50,4)	19,7 (50,6)	19,6 (50,8)	19,5 (51,0)
2	Yarn twist K, kr/m.	358	425	494	566	635	707
3	Relative tissue density E,	63,9	63,7	63,6	63,4	63,3	63,1

	%						
4	Percentage of uncovered fabric area, *f*, %	36,1	36,3	36,4	36,6	36,7	36,9
5	***lg^f***	1,350	1,354	1,357	1,361	1,364	1.368
6	Air permeability coefficient, *C,* cm /cm^{32} sec.	12,3	12,6	12,9	13,2	13,5	13,6
7	Indicator τ of the degree of pressure at 5 mmHg.	0,796	0,7932	0,7903	0,7875	0,7847	0,7838
8	Air permeability of fabric, cm /cm^{32} sec.	44,3	45,2	46.02	46,88	47,73	48,02

A methodology for analysing an experimental fabric sample has been developed. Where the following parameters of the square plain weave fabric were determined, such as the relative density of the fabric ***E, the*** percentage of uncovered area of the fabric ***f***, the air permeability coefficient, ***C, the*** index of degree ***τ*** and air permeability. Indices of square weave fabric parameters as a function of yarn twist ratio in the metric system **an were** obtained. Table 3.7 shows the results of calculation of the parameters of the square structure fabric depending on the yarn twist ratio. Table 3.7 shows that as the yarn twist ratio increases in the metric system, the percentage of uncovered area of the fabric, the air permeability coefficient and the air permeability of the square structure fabric increase, while the relative density of the fabric and the index of the degree of pressure decrease. Comparative studies, square and non-square structure fabrics were carried out. Various criteria and methods have been developed to assess the intensity of fabric production on weaving machines to justify the assortment possibilities of different types of weaving machines. One such criterion is the filling factor of the fabric with fibrous material, taking into account the phase order of the fabric structure. Proceeding from the purpose of the fabric, the parameters of the fabric structure are chosen: the type of raw materials of warp and weft, coefficients depending on the type of raw materials used, thread diameter before weaving, the coefficient of the ratio of thread diameters, the type of weave, the rapport on warp and weft, the order of the phase of the structure (**PFS**), which is characterised by the number and mutual location of levels, i.e. lines drawn through the centres of any group of threads of the same system. As the majority of fabrics for clothing purposes have a structure close to square, therefore the PPS should be from 4 to 6. In these fabrics with equal linear density of warp and weft yarns, the density of the warp fabric usually exceeds the density of the weft fabric by ***n*** times (***n*** up to 1.5 times).

For square fabrics, the surface density of the fabric, the warp and weft yarn number, and the warp and weft densities of the fabrics have the same values. Consequently, the surface density of square and non-square fabrics as well as the linear yarn density of warp and weft are the same. Without disturbing this balance it is possible to obtain a similar non-square fabric with unchanged warp and weft yarn numbers and variable warp and weft densities. If the fabric density of one yarn system is increased by ***n*** times, the fabric density of the other yarn system must be reduced by ***n*** times. For a similar fabric of non-square structure we have at n = 1,45; R_u = ***150*** n/dm; P_o = 320 n/dm; N_o = N_y = 20 or T_o = T_u = 50 tex.

Table 3.8

Parameters of fabrics of square and non-square structure

№	Name	Fabric density		Yarn number (linear density, tex)	
		on the basis of yarn/dm. P_o	in weft yarn/dm. R_u	on the basis of N_o	on duck N_y
1	Fabric of square structure	220	220	20 (50)	20 (50)
2	Number of increases or decreases, ***n*** times.	x 1,45	: 1,45	-	-
3	Non-square fabric	320	150	20 (50)	20 (50)

Table 3.9

Influence of weave on the filling factor of square fabrics

№	Fabric weaves	Weave report		Fill factor		
		R_o basis	ducks R_y	foundation of K_{no}	K_{nu} ducks	gkan K_t
1.	Binding 1/1	2	2	1,228	1,228	1,508
2.	Weave 1/2	3	3	1,023	1,023	1,046
3.	Weave 1/3	4	4	0,937	0,937	0,878
4.	Weave 1/4	5	5	0,859	0,859	0,738

When selecting the above parameters, it should be taken into account that in fabrics with a **PFS** (order of construction phase) lower than 5, the weft yarns protrude to the surface. In such fabrics, the weft density is maximised. The weft filling factor K_{hy} is close to 1. $K_{ho} < K_{hy}$, a_y (weft yield) > a_o (warp yield). In fabrics with a **PFS** above 5, the warp yarns protrude to the surface. The warp density is close to the maximum. The warp fill factor K_{ho} is close to 1. $K_{hy} < K_{ho}$, $a_o > a_y$. It is practically impossible to weave fabrics with the maximum density of both yarn systems on the loom. Therefore, fabrics with lower (actual) densities of both warp and weft systems are produced. The ratio of the actual density to the maximum density is characterised by the fibre filling of the fabric. The fibre fill factor takes into account the fabric density, the linear yarn density and the type of weave of the fabric. The fibre fill factor shows the tension of the fabric production on the weaving machine. Let's calculate the fibre fill factor and present the following

results in Table 3.9 for square weave fabric and in Table 3.10 for non-square weave fabric. The analysis of Table 3.9 and Table 3.10 shows that the fibre filling ratio is different when the weave of the fabric is changed. The process of producing square and non-square fabrics on the loom with weave 1/1 is the most stressful, because the fibre fill factor is greater than one, i.e. *KT>1*.

Table3.10.

Influence of weave on the filling factor of non-square fabrics

№	Fabric weaves	Weave report		Fill factor		
		R_o basis	ducks R_y	foundation of K_{no}	K_{nu} ducks	K_t fabric
1.	Binding 1/1	2	2	1,786	0,837	1,495
2.	Weave 1/2	3	3	1,488	0,698	1,039
3.	Weave 1/3	4	4	1,339	0,628	0,841
4.	Weave 1/4	5	5	1,245	0,586	0,730

The reduction of the weft density from 220 n/dm (square weave) to a weft density of 150 n/dm (non-square weave) results in a 32% reduction in the tension of the weaving process. Here we should expect a reduction in thread breakage and, as a consequence, an increase in the productivity of the weaving machines. In addition, for square and non-square fabrics, the fibre filling factor decreases by 51% with increasing weave ratio in warp and weft.

The fourth chapter is devoted to comparative studies of standard and developed innovative suit fabrics on the basis of their structural characteristics and physical and mechanical properties. The optimal combination of fibre mixtures such as artificial bamboo wool in the fabric is substantiated. Suit fabrics are investigated by the following parameters as surface density, air permeability, abrasion resistance, tensile strength, elongation and moisture absorption. The advantage in structural characteristics and physical and mechanical properties of the developed innovative suit fabric in the weft of the blend of 90% bamboo and 10% wool is shown. The parameters of fabrics of square structure are given. Fabric structure is usually understood as a mutual arrangement of warp and weft yarns in it, caused by their interaction. Interaction forces between threads in the fabric are created in the process of its formation on the weaving machine and determine the mutual arrangement of threads in the fabric. The mutual arrangement of the threads in the fabric and therefore depends on many factors: the type of raw material used; the diameters of warp and weft threads and their ratios; the warp and weft densities and their ratios; the type of weave of the threads in the fabric; the tension of the warp and weft threads and the tension ratios; the technological parameters of threading and fabric production.

The type of raw material for the designed fabric is selected taking into account the purpose of the fabric and the requirements to it. The properties of the yarns used in warp and weft largely determine the properties of the fabric made from them. A change in the type of raw material in at least one system of yarns in the warp or weft of the fabric has a significant impact on the technological parameters of its production, the structure of the fabric and its properties. The diameters of warp and weft yarns used for fabric production have a significant influence on the technological parameters of fabric production, not its structure and properties. When designing a fabric, yarn diameters are determined depending on the purpose of the fabric and the requirements to it. Increasing the diameter of weft yarns increases the breaking load and elongation of the fabric in the weft direction, the working out of the main yarns and reduces the working out of the weft. Consequently, the ratio of warp and weft yarn diameters has a great influence on the parameters, structure and properties of the fabric. Fabric density of warp and weft, and their ratios have a great influence on the structure and properties of fabrics. The change of fabric density by weft, other things being equal, causes the change of technological parameters of production, structure and properties of fabrics. In particular, an increase in weft density leads to an increase in warp tension and a decrease in warp working and fabric width. The warp and weft densities depend on the diameter of the yarns used and the type of weave of the yarns in the fabric. The maximum possible density of short overlap weave fabrics is lower than any other weave in long overlap fabrics. Fabrics with the highest possible warp and weft densities are very difficult to weave on the loom. Most of the fabrics produced have a density of one or both yarn systems that is less than the maximum. Therefore, the ratio of the actual fabric density to the maximum density characterises the filling of the fabric with fibrous material, i.e. the tension of the fabric production on the loom. The type of weave has a great influence on the structure and properties of the fabric. In particular, fabrics with short overlaps have a

higher breaking load and warp and weft yarn working out than fabrics of other types of weaves with long overlaps, and the working out of fabrics with short overlaps is accompanied by high tension. Among the technological parameters that have a significant influence on the structure and properties of the fabric are the tension of warp and weft yarns and their ratios, which change the arrangement of yarns in the fabric, and consequently, the working of yarns in the fabric, the breaking load of the fabric. Another main parameter of threading on the machine is the value of the backstitch, which determines the value of additional warp tension at the moment of forming (surfing) the fabric. As the backstitch increases, the tension of the warp threads at the moment of surf increases, which will lead to changes in the structure of the fabric - density, yarn work in the fabric, breaking load and elongation of the fabric. Hence, with the change in the filling tension and the amount of machine scoring, the structure and properties of the fabrics can be changed. In addition, the structure and properties of fabrics are influenced by the scalo position, height and depth of shed, position of spar, etc. Here we have set a task to study the influence of fabric structure parameters, when weaving woven fabrics with different weaves with short, medium and long overlaps, so it is to be expected that the yarns have different stress states both when the fabric is formed on the loom and after the fabric is removed from the loom. Let us consider the effect on geometric and maximum warp and weft densities for a fabric with warp and weft counts in the fabric between 2 and 5, and with average values of
transition of warp and weft yarns in the fabric. The calculations are as follows,
of a known methodology.

1. The diameter of the thread in the fabric:

warp to weft

$$d_o = 0{,}03162\, \eta_o C_o \sqrt{T_o} \quad (4.1) \qquad d_y = 0{,}03162\, \eta_y C_y \sqrt{T_y}, \quad (4.2)$$

average thread diameter $d_{cp} = \frac{d_o + d_y}{2}$

2. Yarn diameter in the fabric as a function of the diameter ratio and the average yarn diameter

warp to weft

$$d'_o = \frac{2K_d d_{cp}}{K_d + 1} \quad (4.3) \qquad d'_y = \frac{2d_{cp}}{K_d + 1} \quad (4.4)$$

3. Limit density of the fabric

warp to weft

$$P_o = \frac{100}{d_o} \quad (4.5) \qquad\qquad P_y = \frac{100}{d_y} \quad (4.6)$$

4. Maximum fabric density

warp to weft

$$P'_o = \frac{100}{l_o} \quad (4.7) \qquad\qquad P'_y = \frac{100}{l_y}. \quad (4.8)$$

5. Height of the filament bending waves

warp to weft

$$h_o = d_{cp} \cdot K_{ho} \quad (4.9) \qquad\qquad h_y = d_{cp} \cdot K_{hy} \quad (4.10)$$

6.Geometric density of the fabric for short overlap weaves

warp to weft

$$l_o = \sqrt{(d_o + d_y)^2 - h_o^2} \quad (4.11) \qquad\qquad l_y = \sqrt{(d_o + d_y)^2 - h_y^2} \quad (4.12)$$

7. Geometric mean weave density for long overlap weaves

$$l_{ocp} = \frac{t_y \cdot \sqrt{(d_o + d_y)^2 - h_o^2} + (R_o - t_y) \cdot d_o}{R_o} \quad (4.13)$$

on the basis of

$$l_{ycp} = \frac{t_o \cdot \sqrt{(d_o + d_y)^2 - h_y^2} + (R_y - t_o) \cdot d_y}{R_y} \quad (4.14)$$

on duck

where: t_o, t_y - number of transitions of main threads and respectively number of weft threads transitions from one side of the fabric to the other side of the fabric within the fabric rapport per one thread; R_o, R_y - weave rapport of the fabric on the warp and weft.

It follows from the formula that the maximum values of geometric density correspond with the weave ratio of the fabric equal to the number of transitions and these values of geometric density decrease with increasing difference between the weave ratio of the fabric and the number of transitions of main and weft yarns.

Tables 4.1 to 4.4 show the effect of the coefficient determining the height of yarn bending waves on the geometric and maximum warp and weft densities for a fabric with variable warp and weft rapport in the fabric from 2 to 5, and the average values of warp and weft yarn transitions in the fabric.

Table 4.1.

Values of geometric and maximum warp and weft densities for a 1/1 weave fabric.

Order of phase structure of the	Coefficient determining the	Height of filament bending	The geometric density of the	Maximum density, yarn/dm

fabric weave 1/1	height of bending waves of yarns		waves, mm.		fabric is mm.			
	based on K_{ho}	on the duck K_{hy}	based on, h_o	on duck, h_y	on the basis of, l_o	on duck, l_y	on the basis of, P_o	by duck, R_u
Marginal	0,27	1,73	0,069	0,445	0,509	0,257	197	389
III	0,5	1,5	0,128	0,386	0,498	0,339	205	295
IV	0,75	1,25	0,193	0,321	0,476	0,401	210	249
V	1	1	0,257	0,257	0,445	0,445	225	225
VI	1,25	0,75	0,321	0,193	0,401	0,476	249	210
VII	1,5	0,5	0,386	0,128	0,339	0,498	295	205
Marginal	1,73	0,27	0,445	0,069	0,257	0,509	389	197

Table 4.2.

Geometric and maximum warp and weft densities for weave 1/2.

Fabric structure phase order weave 1/2	Coefficient determining the height of bending waves of yarns		Height of filament bending waves, mm.		The geometric density of the fabric is mm.		Maximum density, yarn/dm	
	based on K_{ho}	on the duck K_{hy}	based on, h_o	Duck h_y	On the basis, l_o	On the duck, l_y	on the basis of, P_o	by duck, R_u
Marginal	0,27	1,73	0,069	0,445	0,425	0,257	253	389
III	0,5	1,5	0,128	0,386	0,418	0,312	239	321
IV	0,75	1,25	0,193	0,321	0,403	0,353	248	283
V	1	1	0,257	0,257	0,382	0,382	262	262
VI	1,25	0,75	0,321	0,193	0,353	0,403	283	248
VII	1,5	0,5	0,386	0,128	0,312	0,418	321	239
Marginal	1,73	0,27	0,445	0,069	0,257	0,425	389	253

Table 4.3.

Values of geometric and maximum warp and weft densities for 1/3 weave fabrics.

Fabric structure phase order weave 1/3	Coefficient determining the height of bending waves of yarns		Height of filament bending waves, mm.		The geometric density of the fabric is mm.		Maximum density, yarn/dm	
	based on K_{ho}	On the duck K_{hy}	on the basics^e, h_o	on duck y^h	on the basis of, l_o	On the duck, l_y	on the basis of, P_o	by duck, R_u
Marginal	0,27	1,73	0,069	0,445	0,383	0,257	261	389
III	0,5	1,5	0,128	0,386	0,378	0,298	265	336
IV	0,75	1,25	0,193	0,321	0,367	0,329	273	304
V	1	1	0,257	0,257	0,351	0,351	285	285
VI	1,25	0,75	0,321	0,193	0,329	0,367	304	273

VII	1,5	0,5	0,386	0,128	0,298	0,378	336	265
Marginal	1,73	0,27	0,445	0,069	0,257	0,383	389	261

Table 4.4.

Geometric and maximum warp and weft densities for 1/4 weave fabrics.

Weave 1/4 fabric construction phase order	Coefficient determining the height of bending waves of yarns		Height of filament bending waves, mm.		The geometric density of the fabric is mm.		Maximum density, yarn/dm.	
	based on K_{ho}	on the duck K_{hy}	based on, h_o	on duck, h_y	on the basis of, l_o	on duck, l_y	on the basis of, P_o	by duck, R_u
Marginal	0,27	1,73	0,069	0,445	0,358	0,257	279	389
III	0,5	1,5	0,128	0,386	0,353	0,290	283	345
IV	0,75	1,25	0,193	0,321	0,345	0,315	299	317
V	1	1	0,257	0,257	0,332	0,332	301	301
VI	1,25	0,75	0,321	0,193	0,315	0,345	317	299
VII	1,5	0,5	0,386	0,128	0,290	0,353	345	283
Marginal	1,73	0,27	0,445	0,069	0,257	0,358	389	279

Fig. 4.1 shows the graphs of dependence of yarns bending wave height on the coefficient determining the height of yarns bending waves on the warp. Fig.4.2 shows the graphs of dependence of geometric density of fabric on warp and weft on the coefficient determining the height of the warp bending waves for weave 1/1. Fig. 4.3 shows the graphs of dependence of the maximum density of the fabric on the basis and on the weft on the coefficient determining the height of the thread bending waves on the basis for weave 1/1. In the work similarly plots of dependence of geometric density of fabric on the base and on the weft and the maximum density of fabric on the base and on the weft on the coefficient determining the height of waves of bending of yarns on the base for weaves 1/2, 1/3 and 1/4 are constructed. The peculiarity of these regularities is that they have constant qualitative indices and variable quantitative indices. It can be noted that the graphs of dependence of the height of thread bending waves on the coefficient determining the height of thread bending waves on the warp for all variants of weaves remained unchanged.

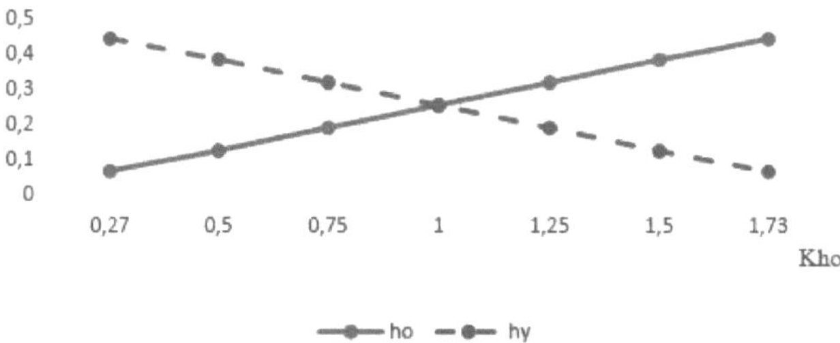

Fig.4.1.Dependence of the height of filament bending waves on the coefficient determining the height of filament bending waves along the warp.

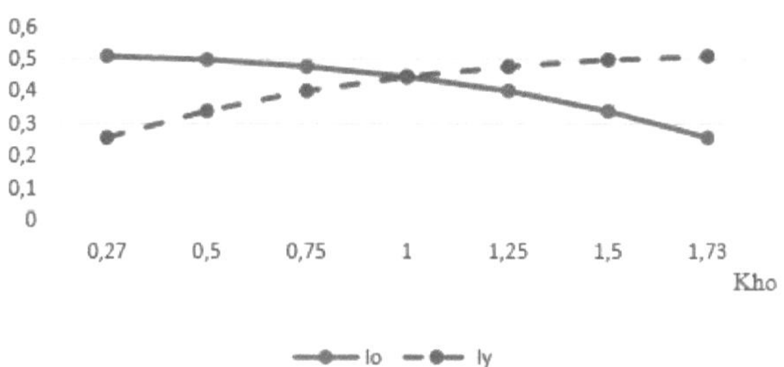

Fig.4.2 Dependence of geometrical density of fabric on warp and weft on the coefficient determining the height of warp bending waves for weave 1/1.

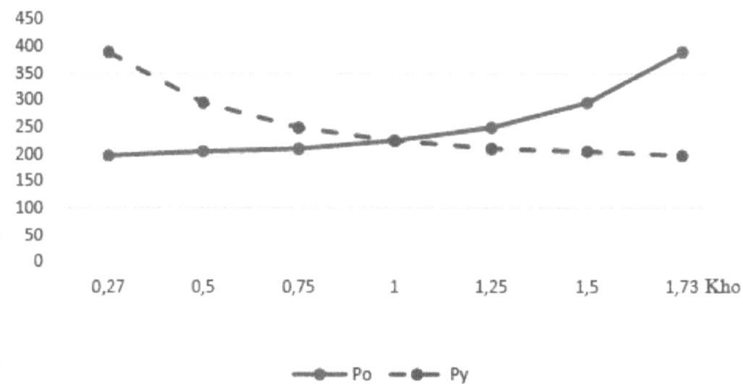

Fig. 4.3. Dependence of the maximum density of the fabric on the warp and

weft
on the coefficient determining the height of the warp bending waves for weave 1/1.

From the analysis of Tables 4.1-4.4 and Figs. 4.1-4.3 it follows that at variable warp and weft rapport and at constant value of the number of yarn transitions in the fabric: the maximum warp and weft density decreases; the geometric warp and weft density increases; the height of the warp and weft bending wave remains unchanged. The influence of yarn twist on the surface density of square structure garment fabrics is also investigated; the filling calculation of square structure fabrics depending on the weave is given. The coefficient of fabric cohesion (**CCT**) and coefficient of fabric filling (**CST**) of square structure depending on weave and yarn twist are determined. The intensity of the production process of square weaving on the loom with weave 1/1 (short overlaps) is higher with respect to weaves 1/2,1/3 and 1/4 (long overlaps). The fabric surface density, filling factor (FFR) and cohesion factor (**CFT**) of a square weave increases with increasing yarn twist. Weaving machines are selected depending on their assortment capacity taking into account high productivity and high quality of fabrics. When determining the assortment possibility of the looms the possibility of fabric production is found out depending on: the kind of warp and weft, types of weft (monochrome, multicolour, etc.); linear density of threads in the fabric; obtaining the necessary weave; obtaining the required width of fabrics; obtaining a certain density of threads in the fabric; the intensity of the process of fabric production. Tension of fabric production process depends on weave, density and thickness of fabric threads and can be characterised by the coefficient of fabric filling with fibrous material or the coefficient of fabric binding. With an increase in the filling factor or the fabric's connectivity, the weaving process becomes more tense and the weaving is more stressful and more difficult. The loom is selected in the following sequence: the cohesion coefficient (**CCT**), the fibre fill factor (**CFF**) are calculated and the loom type is selected; the working width of the loom is selected based on the reed filling width, which ensures the production of a rough fabric, which after finishing will meet the width requirements of GOST "Finished fabric widths". The calculations are given below for fabrics of square structure, i.e. fabric densities and linear yarn densities of warp and weft are the same.

Fabric filling ratio (**FFR**) with warp yarns

$$K_{H_o} = \frac{P_o \cdot (d_o \cdot R_o + d_y \cdot t_y)}{R_o \cdot 10}$$

Filling factor of the fabric with weft yarns

$$K_{H_y} = \frac{P_y(d_y \cdot R_y + d_o \cdot t_o)}{R_y \cdot 10}$$

where: R_o, R_u - warp and weft fabric density, thread / mm; d_o, d_y - warp and weft thread diameters, mm; R_o, R_y - warp and weft weave rapports; t_o - number of warp thread transitions from one side of the fabric to another within the rapport per thread; t_y - number of weft thread transitions from one side of the fabric to another within the rapport per thread.

Filament diameters are determined by the formula

$$d = 0.0316 \cdot c\sqrt{T}$$

where: c - coefficient depending on the type of yarn (selected according to the reference book);

T - linear yarn density, tex.

Filling factor of woollen fabric

$$K_{NT} = K - K_{nu}$$

Tissue cohesion coefficient (**TCC**) is calculated by the formula

$$C = \frac{P_o \cdot P_y \cdot \bar{T}}{F \cdot 1000}$$

The average linear density of warp and weft yarns is calculated by the formula

$$T = \frac{2 \cdot T_o \cdot T_y}{T_o + T_y}$$

where : T_o, T_u - respectively linear densities of warp and weft yarns;

F - average weave coefficient, calculated by the formula

$$F = \frac{2R_oR_y}{r_o + r_y}$$

Where: d_o - number of warp and weft links in the warp direction within the rapport; d_y - the same in the weft direction.

For weaving fabric 1/1 $R_o = R_y = 2$, $t_o = t_y = 2$, $ч_o = ч_y = 2$, $F = 2$.

For weaving fabric 1/2 $R_o = R_y = 3$ $t_o = t_y = 2$, $ч_o = ч_y = 3$, $F = 3$.

For weaving fabric 1/3 $R_o = R_y = 4$ $t_o = t_y = 2$, $ч_o = ч_y = 4$, $F = 4$.

For weaving fabric 1/4 $R_o = R_y = 5$ $t_o = t_y = 2$, $ч_o = ч_y = 5$, $F = 5$.

The weaving machine type is selected taking into account the calculation of K_{NT}(**KST**) and ***C*** (**KST**) as well as T_o, T_y, P_o, P_y and the weave ratio. The working width of the weaving machine is selected according to the width of the woven fabric. The width of the wool fabric produced on the machine must be such that after finishing the width of the finished fabric corresponds to GOST "Widths of the finished fabric". Calculation of reed number

$$N_б = \frac{P\ (1 - \frac{a}{100})}{Z_Ф}$$

z_φ - *the* number of warp threads that are threaded through the reed tooth.
Weight of warp yarns in square fabrics

$$M_о = \frac{T\ \cdot \mathrm{P}}{10^6 \cdot (1 - \frac{a}{100})}$$

where: T - linear density of yarns, tex; ***P*** - density of square fabric; ***a*** - yarn processing.
Weight of weft yarns in square weave fabrics

$$M_у = \frac{T\ \cdot \mathrm{P}}{10^6 \cdot (1 - \frac{a}{100})}$$

Weight per linear metre of square structure fabric

$$M_{пм} = M_о + M_у \quad \text{gr/m}$$

Surface density of square structure tissue

$$M_{М^2} = \frac{M_{пм}}{В_с} \quad \text{gr/m}^2$$

Let's carry out the filling calculation of the square fabric depending on the weave 1/1,1/2,1/3,1/4 (Fig.4.11) and yarn twist. The results are given in table 4.5.

Table 4.5.

Results of the filling calculation of the square structure fabric

№	Name	Yarn twist ratio in metric system $_{en}$, (in tex system $_{at}$)					
		50 (16)	60 (19)	70 (22,1)	80 (25,3)	90 (28,4)	100(31,6)
1	Average metric yarn number ***N*** (linear density ***T***, tex) in the fabric	20 (50)	19,9 (50,2)	19,8 (50,4)	19,7 (50,6)	19,6 (50,8)	19,5 (51,0)
2	Fabric density yarn/dm.	220	220	220	220	220	220
3	Yarn diameter	0,279	0,2799	0,280	0,281	0,2815	0,282
4	**CST** **CCP** fabric weave 1/1	1,507 12,10	1,517 12,15	1,518 12,20	1,529 12,25	1,534 12,30	1,540 12,40
5	**CST**	1,046	1,053	1,054	1,062	1,065	1,069

	CCP weave 1/2	8,07	8,10	8,13	8,16	8,20	8,30
6	**CST** **CCP** fabric weave 1/3	0,848 6,05	0,852 6,08	0,854 6,10	0,860 6,12	0,863 6,15	0,866 6,20
7	**CST** **CCP** 1/4 weave	0,738 4,84	0,743 4,86	0,744 4,88	0,749 4,90	0,752 4,92	0,755 5,00
8	Yarn yield in fabric, %	9	9	9	9	9	9
9	Weight of weft warp yarns, g/m.	120	121	122	122	123	124
10	Surface density of fabric, gr/m^2	240	242	244	244	246	248

In Fig. 4.4 and Fig. 4.5 show the effect of fabric weave on fabric fill factor and fabric coefficient of connectivity. Fig. 4.6 and Fig. 4.7 show the effect of yarn twist ratio on fabric filling ratio and fabric cohesion ratio. Fig. 4.8 shows the effect of yarn twist ratio on fabric surface density.

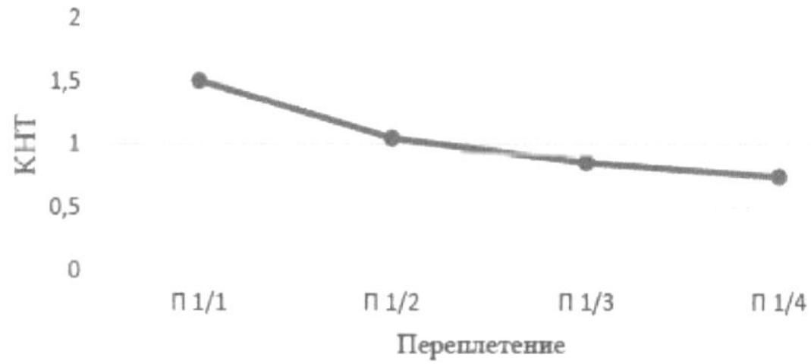

Figure 4.4. Effect of fabric weave on fabric fill factor.

Figure 4.5. Influence of fabric weave on the fabric's coefficient of connectivity.

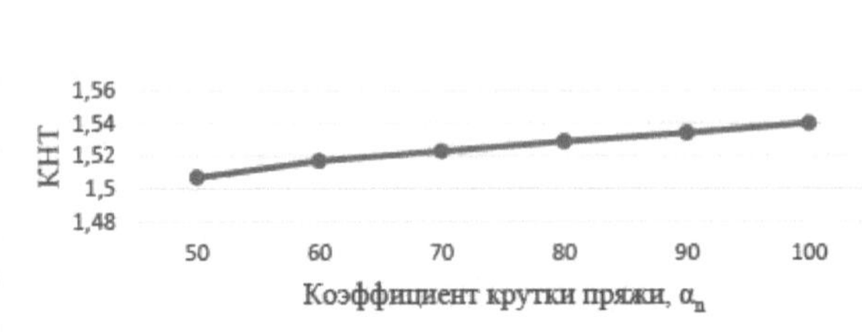

Figure 4.6. Influence of yarn twist ratio on fabric filling factor

.

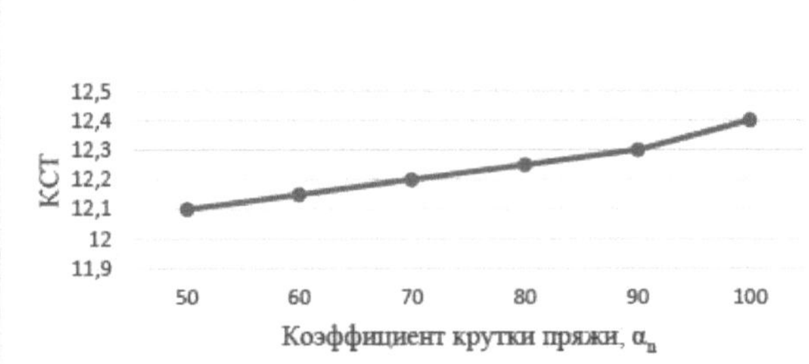

Figure 4.7. Influence of yarn twist ratio on fabric coefficient of connectivity.

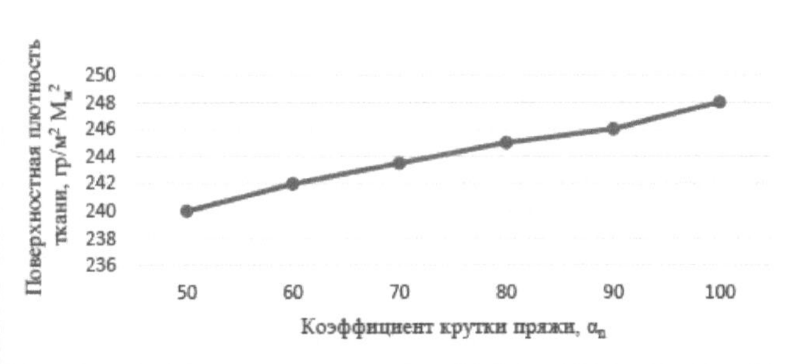

Figure 4.8. Influence of yarn twist ratio *αn* on the surface density of the fabric.

Analyses of Table 4.5 and Figs. 4.4 - 4.8 show that the fibre filling ratio (FFR) and the fabric cohesion coefficient (**FCR**) have different values when the weave changes from short overlaps to long overlaps. The most strenuous weaving process on the loom occurs in the 1/1 weave (short overlaps). Increasing the yarn twist leads to an increase in the filling factor for 1/1,1/2,1/3,1/4 weaves and also to an increase in the surface density of the fabric. The development and production of innovative suit fabric has been carried out. Woollen fabrics with different percentage of wool are used in the production of garments. Woollen fabrics in the general assortment occupy a small specific weight (about 10 %), but by the number of articles are very diverse, they have a narrower purpose, as well as longer service life, are characterised by high elasticity, low wrinkling, good heat-protective properties and form stability. Woollen fabrics are produced pure woollen, woollen and semi woollen. Pure wool fabrics contain 100% wool or have in their composition up to 5% of other (usually chemical) fibres introduced to give certain external effects (lustre, grey hair, coloured dregs). Pure woollen fabrics made of fine wool are the most valuable, they have the best heat-protective properties, beautiful appearance, wear-resistant. Woollen fabrics should contain at least 70% of woollen fibres. Semi woollen fabrics differ in the content of wool and additionally introduced fibres, types of these fibres, the way of their introduction. The content of wool in semi-woolen fabrics can be from 10 to 70%, and the rest of the components may be high-tech chemical fibres with unique properties (bamboo, aramid, polyethylene, carbon, etc.). Of greatest interest is bamboo fibre. This innovative material has many unique features, primarily for supporters of a healthy lifestyle, and their reviews of bamboo fibre and fabrics made of it are unanimous and positive. Bamboo is a herbaceous plant of hot climates, the main feature of which is a high growth rate and unpretentiousness. Unlike cotton, it does not deplete the soil and does not require treatment with chemicals during cultivation. In addition, this high-growing grass has many beneficial properties and its composition contains many valuable substances that promote health.

Artificial fibre from bamboo began to produce in 2000, and a few years later products from this fabric began to actively conquer the market.

Therefore, it is reasonable to use bamboo fibres in a mixture with wool fibres for the production of suit fabrics. Two samples of fabrics were chosen for research of suit fabrics properties - standard fabric of article 23195, where the main and weft yarns are a mixture of wool and lavsan, and innovative fabric with insertion of a mixture of bamboo and wool in the weft and a mixture of wool and lavsan in the base. In accordance with the developed research programme the following structural characteristics and physical and mechanical properties of fabrics were determined: surface density of fabrics, g/m^2 ; air permeability, dm^3 /(dm^2 s); abrasion resistance, cycles; breaking strength, N; elongation, %; moisture absorption, %. To determine the structural characteristics and surface density, the investigated fabric samples were pre-conditioned under normal conditions in accordance with the established requirements. Structural characteristics and physical-mechanical properties of fabrics were determined on the devices of the laboratory "CENTEXUZ" at TITLP.

Table 4.6 shows physical and mechanical properties of the proposed fabric and tricot fabric with lavsan article 23195, Table 4.7 shows physical and mechanical parameters of yarns depending on the content of artificial bamboo fibres in wool for weft yarn, and Table 4.8 shows comparative characteristics of properties of bamboo fibres with properties of some other fibres.

Table 4.6

Physico-mechanical parameters of the fabric.

№	Fabric sample	Yarn blend composition, %		Surface density of fabric, gr/m^2	Air permeability, cm /cm^{32} sec.	Abrasion resistance, cycles
		basis	ducks			
1	Existing fabric art. 23195	50% Wool 50% Polyester	50% Wool 50% Polyester	190	11	4060
2	New fabric	50% Wool 50% Polyester	10% Wool 90 Bamboo	180	48	4100

Table 4.7

Physical and mechanical properties of yarns.

№	Percentage of bamboo insertion in wool for weft yarns	Breaking strength, N	Elongation, %	Moisture absorption, %
1	Bamboo: wool-90:10%	260	24	70
2	Bamboo: wool-70:30%	220	21	68
3	Bamboo: wool-50:50%	200	20	67

Comparative characterisation of the properties of bamboo fibre with those of some other fibres

Name of indicators	Bamboo	Wool	Viscose	Cotton	Polyester
Relative strength, cH/tex	40-42	11-13	22-26	20-24	55-60
Elongation, %	14-16	25-30	20-25	7-9	25-30
Relative wet strength, cH/tex	34-38	8-10	10-15	26-30	54-58
Elongation in wet condition, %	16-18	27-35	25-30	12-14	25-30

Advantages of bamboo: high breathability, which is achieved due to the porous structure; high durability - both in dry and wet form; hygroscopicity of bamboo products, which quickly absorb moisture; thermal insulation - in winter, the fabric will not let the body freeze, as it will accumulate heat; resistance to UV rays - do not deteriorate under the influence of sunlight; resistance to odours; pliability to dyeing; resistance to wear - has a long service life, as they do not deteriorate under the influence of external factors; softness; drapability - perfectly forms folds without creases; antibactericidal properties - bamboo effectively destroys pathogenic microorganisms that have penetrated its surface; ecological purity; hypoallergenic - bamboo does not cause allergic reactions, so it is suitable for children and people with sensitive skin; easy care. Analysis of tables shows that the surface density of the new fabric is 5% lower than the standard fabric. It is caused by the fact that artificial bamboo fibres have a lower specific weight of fibres due to high porosity. In addition, the fabric's abrasion resistance is within the standard for this fabric. The new fabric is almost four times more breathable than the existing fabric. From the physical and mechanical parameters of the yarn it follows that the breaking strength increases depending on the percentage of bamboo inclusion in the wool for the yarn. Consequently, the proposed material allows to produce hygienic fabrics of costume assortment. Moreover, the proposed material is economical due to the use of blended yarn. Optimisation of the weaving process in the production of innovative suit fabric has also been carried out. Preliminary study of the object of research showed that the purpose of optimisation of the weaving process is to achieve the maximum level of equipment productivity. Reducing the level of thread breakage in the weaving process to the greatest extent (at a constant speed of the main shaft of the machine) contributes to the achievement of this goal. This allows us to choose the minimum of the main thread breakage as the main criterion for optimising the weaving process. On the one hand, it determines such indicators as weaver's workload, equipment utilisation rate, which significantly influence the labour and equipment productivity. On the other hand, the thread breakage index can unambiguously determine the conditions of fabric formation on the weaving machine depending on the technological parameters of its threading and establish the most favourable conditions of fabric formation according to the minimum value of thread breakage. Thus, the adopted optimisation criterion allows to characterise the efficiency of the research object with the greatest completeness. Three main parameters were chosen, i.e. three main independent variables: x_1 - warp tension, cN; x_2 - the size of the scalp, mm; x_3 - the position of the scalp in relation to the sternum in height, mm. It is established that warp breakage at low values of the warp setting tension increases due to the increase of the surge. Then, as the warp tension increases, the breakage rate decreases and increases again as the warp tension increases further due to warp overstressing. The size of the backstop affects the conditions of the weft fringe and, as a consequence, the size of the fringe. As the backstop increases, the weft yarn run-in conditions improve and the size of the surf strip decreases. Therefore, when producing fabrics with a higher filling ratio, a larger backstop is recommended. As the backstop decreases, the size of the bump increases and a dynamic shock phenomenon occurs, which can increase the breakage of warp yarns. The degree of movement of the weft on the warp yarns at the moment of surf depends on the tension of the main yarns in the upper and lower parts of the shed. When the tension of the shed branches is different, more favourable conditions are created for the weft yarn surfing, i.e. the conditions of fabric formation are improved and warp yarn breakage is reduced. On the other hand, large deviations from the level of symmetrical shed can create a weakening of tension in one branch and increase it in the other, i.e. lead to breakage of weakly tensioned warp yarns. The difference in tension between the upper and lower branches of the shed can be controlled by the height of the cliff. The selected factors meet all the requirements of the theory of mathematical planning of experiment: there is no interchangeability of factors, they can be measured by available means, they can be varied within a sufficiently wide range of minimum and maximum values and taken with the necessary accuracy. As for the other technological parameters of loom fuelling, all of them were constant during the experiment. Since the weaving process is non-stationary in time, and a large number of experiments distort the results due to different disturbances of the process, randomisation of experiments was applied in the planning of the experiment. In this work, the central composite method of planning the experiment of the second order was adopted, which gives the possibility of detailed study, description and optimisation of the weaving process in the investigated area of optimisation. The selection of intervals and values of factors for five levels of variation was carried out taking into account the technological possibilities of machine filling (Table 4.9).

Table 4.9

Levels of variation in factors.

Factors	Levels of variation					Interval
	-1,682	-1,0	0	+1,0	+1,682	
χ_1-filling warp tension, cN	14	16	19	22	24	3
X_2-value of offset, mm	10	14	20	26	30	6

xz-position of the scalp relative to the sternum, mm	-25	-15	0	+15	+25	15

The experiment performed on the selected matrix allows us to obtain the a second-order mathematical model describing the influence of factors x1, x2, x3 on the selected optimisation parameters of the following form

$$y = в_0 + в_1 x_1 + в_2 x_2 + в_3 x_3 + в_{12} x_1 x_2 + в_{23} x_2 x_3 + в_{13} x_1 x_3 + в_{11} x_1^2 + в_{22} x_2^2 + в_{33} x_3^2$$

wherew, Bi, Bij, Bu - coefficients; regression;

in$_0$ is a free member;

Bi=1, 2, 3 - regression coefficients at linear terms;

Bij=1, 2, 3 - coefficients in the interaction of factors;

Bii= - regression coefficients of squared terms.

Calculation of regression coefficients

$$в_0 = g_1 \sum_{u=1}^{N} \overline{y}_u - g_2 \sum_{i=1}^{M} \sum_{u=1}^{N} x_{iu}^2 \overline{y}_u$$

$$b_0 = 0{,}1663(4{,}54) - 0{,}0568(10{,}56) = 0{,}155$$

$$в_1 = g_2 \sum_{u=1}^{N} x_{iu} \overline{y}_u$$

$$b_1 = 0{,}0732(-0{,}697) = -0{,}051$$

$$b_2 = 0{,}0732(0{,}0656) = 0{,}005$$

$$b_3 = 0{,}0732(0{,}123) = 0{,}009$$

The pairwise interaction coefficients of the regression equation were determined using the formula:

$$в_{ij} = g_u \sum_{u=1}^{N} x_{iu} \cdot x_{ju} \cdot \overline{y}_u$$

Table 4.10.

RCCE planning matrix

Order of randomisation	Experience number	Factors			Optimisation criterion U warp breakage rate	(Wee^)^2
		X1	X2	X3		
20	1	+	+	+	0,19	0,054
19	2	+	+	-	0,11	0,037
18	3	+	-	+	0,19	0,0064
17	4	+	-	-	0,25	0,0105
16	5	-	+	+	0,28	0,0056
15	6	-	+	-	0,33	0,0065
14	7	-	-	+	0,19	0,0016
13	8	-	-	-	0,39	0,0081
1	9	0	0	0	0,16	0,000025

2	10	0	0	0	0,15	0,000025
3	11	0	0	0	0,17	0,000225
4	12	0	0	0	0,16	0,000025
5	13	0	0	0	0,15	0,000025
6	14	0	0	0	0,15	0,000025
11	15	+1,682	0	0	0,40	0,0072
12	16	-1,682	0	0	0,55	0,0049
9	17	0	+1,682	0	0,23	0,0039
10	18	0	-1,682	0	0,30	0,0042
7	19	0	0	+1,682	0,08	0,0006
8	20	0	0	-1,682	0,11	0,0016

$b_{13} = 0.125(0.272) = 0.034$

$b_{23} = 0.125(0.288) = 0.03\ 6\ 6$

$b_{12} = 0.125(0.064) = 0.008$

The coefficients of the quadratic terms of the regression equation are determined by the formula:

$$e_{ij} = g_6 \sum_{u=1}^{N} x_{iu}^2 \cdot \bar{y}_u + g_6 \sum_{i=1}^{M} \sum_{u=1}^{N} x_{iu}^2 \cdot \bar{y}_u - g_2 \sum_{u=1}^{N} y_u$$

$$b_{11} = 0{,}0625(4{,}62) - 0{,}0069(10{,}56) - 0{,}0568(4{,}54) = 0{,}104$$

$$b_{22} = 0{,}0625(3{,}44) - 0{,}0069(10{,}56) - 0{,}0568(4{,}54) = 0{,}03$$

$$b_{33} = 0{,}0625(2{,}48) - 0{,}0069(10{,}56) - 0{,}0568(4{,}54) = -0{,}03$$

According to the methodology of processing the experimental results, the variance of the output parameter in the experiment or the variance of reproducibility is determined :

$$S^2\{\bar{y}\} = S_ч^2\{y\} = \frac{1}{N_ч - 1} \sum_{U_ч=1}^{N_ч-6} \left(y_{u_ч} - \bar{y}_ч\right)^2$$

$$S^2\{\bar{y}\} = \frac{1}{5} \cdot 0{,}000134 = 0{,}00003 \qquad \bar{y}_ч = 0{,}113$$

$$S^2\{\bar{y}\} = \frac{1}{5} \cdot 0{,}00135 = 0{,}00027 \qquad \bar{y}_ц = 0{,}155$$

Next, the variance of the regression coefficients is determined:

$$S^2\{b_0\} = g_3 \cdot S^2\{\bar{y}\}$$

$$S^2\{b_0\} = 0{,}1663 \cdot 0{,}00027 = 0{,}000045 \qquad S\{b_0\} = 0{,}0067$$

$$S^2\{b_i\} = g_1 \cdot S^2\{\bar{y}\}$$

$$S^2\{b_i\} = 0{,}0732 \cdot 0{,}00027 = 0{,}00002 \qquad S\{b_i\} = 0{,}0044$$

$$S^2\{b_{ij}\} = g_4 \cdot S^2\{\bar{y}\}$$

$$S^2\{b_{ij}\} = 0{,}125 \cdot 0{,}00027 = 0{,}000034 \qquad S\{b_{ij}\} = 0{,}0058$$

$$S^2\{b_{ii}\} = g_7 \cdot S^2\{\bar{y}\}$$

$$S^2\{b_{ii}\} = 0{,}0695 \cdot 0{,}00027 = 0{,}000019 \qquad S\{b_{ii}\} = 0{,}0043$$

To determine the significance of the regression coefficients, Student's criteria are used

$$t_R = \frac{b}{S\{b\}} \qquad t_R\{b_0\} = \frac{0{,}155}{0{,}0067} = 23{,}1$$

$$t_R = \frac{b_i}{S\{b_i\}}$$

$$t_R\{b_1\} = \frac{0{,}051}{0{,}0044} = 11{,}6 \quad t_R\{b_2\} = \frac{0{,}005}{0{,}0044} = 1{,}2 \quad t_R\{b_3\} = \frac{0{,}009}{0{,}0044} = 2.0$$

$$t_R = \frac{b_{ij}}{S\{b_{ij}\}}$$

$$t_R\{b_{12}\} = \frac{0{,}008}{0{,}0044} = 1{,}8 \quad t_R\{b_{13}\} = \frac{0{,}034}{0{,}0058} = 5{,}9 \quad t_R\{b_{23}\} = \frac{0{,}036}{0{,}0058} = 6{,}2$$

$$t_R = \frac{b_{ii}}{S\{b_{ii}\}}$$

$$t_R\{b_{11}\} = \frac{0{,}104}{0{,}0043} = 24{,}2 \quad t_R\{b_{22}\} = \frac{0{,}03}{0{,}0043} = 7{,}0 \quad t_R\{b_{33}\} = \frac{0{,}03}{0{,}0043} = 7{,}0$$

Tabular value Student's criterion

$$t_T\left[P_D = 0{,}95; \qquad fS_y^2 \; 6-1=5\right] = 2{,}571$$

Regression coefficients are significant if $t_R > t_i$. In this case, not all coefficients are significant, in particular the coefficients at X_2 , X_3 and the pairwise interaction coefficient X1X2 have smaller values, that is $t_T > t_R$. Therefore, we discard from further processing. The mathematical model describing the dependence of breakage on the selected factors is as follows

y_R = 0.155 - 0.051Y38E↩1 + 0.104X12 + 0.03X22 - 0.03X_3^2 + 0.034X1X3 + 0.036X2X3 (1)

To test the hypothesis about the adequacy of the obtained model, we use Fisher's criterion, the calculated value of which is compared with the tabulated F_T. If $F_R < F_T$, then with probability P_D the hypothesis of adequacy of the obtained model is not rejected. The calculated values of Fisher's criterion are equal to:

$$F_R = \frac{S_{ad}}{S_E}$$

$$S_E = \sum_{u=1}^{N_u} (y_u - \bar{y})^2$$

$$S_R = \sum_{u=1}^{N} (\bar{y} - y_R)^2$$

$$S_{ad} = S_R - S_E$$

$$S_E = 0{,}00027 \quad S_R = 0{,}0014$$

$$S_{ad} = 0{,}0014 - 0{,}00027 = 0{,}00113$$

$$F_R = \frac{0{,}00113}{0{,}0027} = 4{,}2$$

Since **FR** < **FT**, at confidence probability PD = 0.95, f_r [P_D = 0.95; *fl* = L?TS - 1 = 6 - 1 = 5; *fl* = *N* - *NK3* " - (L?TS - 1) = 20 - 7 - (6 -1) = 8] = 4.82 the hypothesis of adequacy of the obtained models is not rejected, i.e. 4.2 < 4.82. We evaluated the process experiment by means of cuts: **Y** = f (x1) at constant x_2 , x_3 **;** **Y** = f (x_2) at constant x1, x_3 ; y = f (x_3) at constant x1, x_2 . Tables 4.11- 4.13 summarise the results of the cliffness calculations from the input factors.

Calculation results *Y=f* (x1) at constant x2 and x3

№	Constant values of factors	Yarn breakage ***at the*** value of factor x1 variables				
		-1,682	-1	0	+1	+1,682
1	X2=-1, XZ=-1	0,628199	0,38	0,191	0,21	0,342259
2	X2=-1, XZ=0	0,565011	0,34	0,185	0,238	0,393447
3	X2=-1, XZ=1	0,441823	0,24	0,119	0,206	0,384635
4	X2=0, XZ=-1	0,562199	0,314	0,125	0,144	0,276259
5	x2=0, XZ=0	0,535011	0,31	0,155	0,208	0,363447
6	x2=0, XZ=1	0,447823	0,246	0,125	0,212	0,390635
7	x2=1, XZ=-1	0,556199	0,308	0,119	0,138	0,270259
8	x2=1, XZ=0	0,565011	0,34	0,185	0,238	0,393447
9	x2=1, XZ=1	0,513823	0,312	0,191	0,278	0,456635

Table 4.12.

Calculation results *Y=f* (x2) at constant x1 and x3

№	Constant values of factors	Warp breakage *y* variables x2 factor value				
		-1,682	-1	0	+1	+1,682
1	x1=-1, XZ=-1	0,459426	0,38	0,314	0,308	0,338322
2	x1=-1, XZ=0	0,394874	0,34	0,31	0,34	0,394874
3	x1=-1, XZ=1	0,270322	0,24	0,246	0,312	0,391426
4	x1=0, XZ=-1	0,270426	0,191	0,125	0,119	0,149322
5	x1=0, XZ=0	0,239874	0,185	0,155	0,185	0,239874

6	x1=0, xz=1	0,149322	0,119	0,125	0,191	0,270426
7	x1=1, xz=-1	0,289426	0,21	0,144	0,138	0,168322
8	x1=1, xz=0	0,292874	0,238	0,208	0,238	0,292874
9	x1=1, xz=1	0,236322	0,206	0,212	0,278	0,357426
10	x1=0, xz=1.682	0,053152	0,039574	0,070126	0,160678	0,256848

Table 4.13.

Calculation results Y=f(x3) at constant x1 and x2

№	Constant values of factors	Warp breakage *y* variables factor value xz				
		-1,682	-1	0	+1	+1,682
1	x1=-1, x2=-1	0,372866	0,38	0,34	0,24	0,137386
2	x1=-1, x2=0	0,282314	0,314	0,31	0,246	0,167938
3	x1=-1, x2=1	0,251762	0,308	0,34	0,312	0,25849
4	x1=0, x2=-1	0,160678	0,191	0,185	0,119	0,039574
5	x1=0, x2=0	0,070126	0,125	0,155	0,125	0,070126
6	x1=0, x2=1	0,039574	0,119	0,185	0,191	0,160678
7	x1=1, x2=-1	0,15649	0,21	0,238	0,206	0,149762
8	x1=1, x2=0	0,065938	0,144	0,208	0,212	0,180314
9	x1=1, x2=1	0,035386	0,138	0,238	0,278	0,270866

Figs. 1 3 are plotted in Figs.

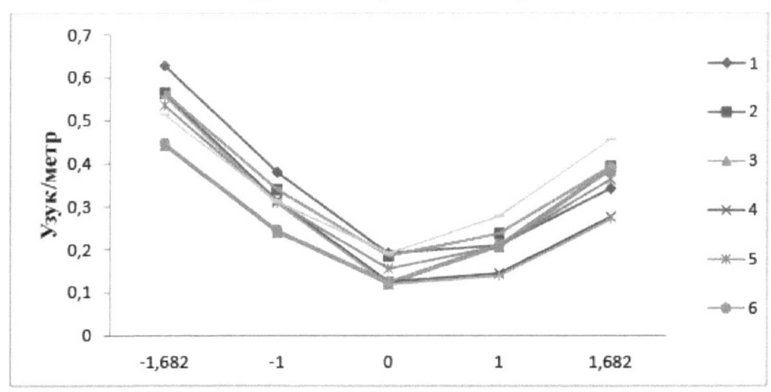

Row1 - X2 = -1; xz = -1 Row2- X2 = -1; xz = 0 Row3 - X2 = -1; xz = 1
Row4 - x_2 = 0; x_3 = -1 Row5- x_2 = 0; x_3 = *0* Row6 - x_2 = *0*; x_3 = 1
Row7- x2 = 1; xz = -1 Row8- x2 = 1; xz = 0 Row9- x2 = 1; xz = 1

Figure 4.9. Influence of warp filling tension on thread breakage.

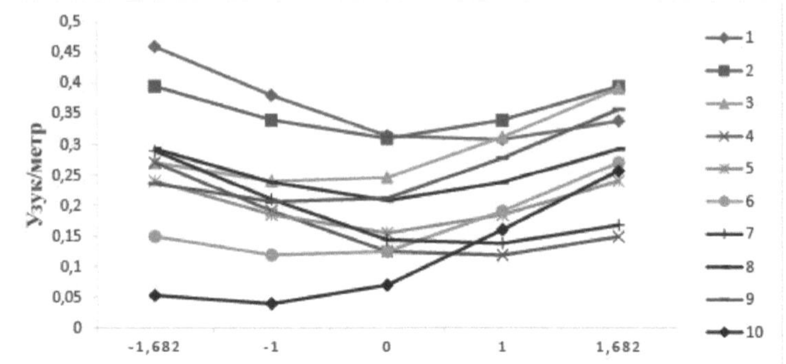

Ряд 1- $x_1 = -1$; $x_3 = -1$; Ряд 2 - $x_1 = -1$; $x_3 = 0$; Ряд 3 - $x_1 = -1$; $x_3 = 1$;
Ряд 4 - $x_1 = 0$; $x_3 = -1$; Ряд 5 - $x_1 = 0$; $x_3 = 0$; Ряд 6 - $x_1 = 0$; $x_3 = 1$;
Ряд 7 - $x_1 = 1$; $x_3 = -1$; Ряд 8 - $x_1 = 1$; $x_3 = 0$; Ряд 9 - $x_1 = 1$; $x_3 = 1$;
Ряд 10 - $x_1=0$; $x_3=1{,}682$.

Fig.4.10. Influence of the size of the gap on thread breakage

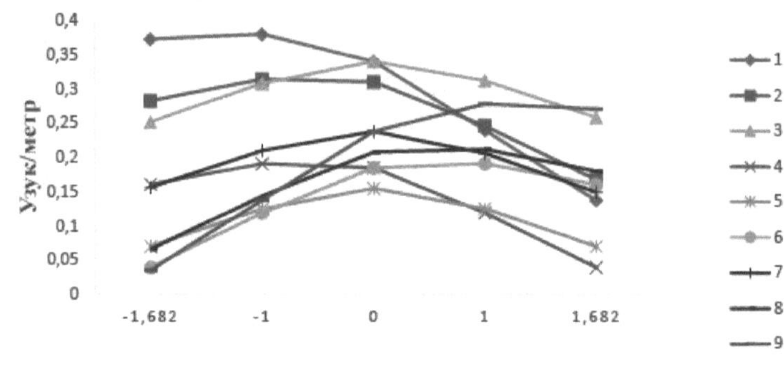

Row 1- x1 = -1; x2 = -1; Row 2- x1 = -1; x2 = 0; Row 3 - x1 = -1; x2 =1 ;
Row 4 - x1 = 0; x2= -1; Row 5- x1 = 0; x2= 0; Row 6 - x1 = 0; x2=1 ;
Row 7 - x1 = 1; x2 = -1; Row 8- x1 = 1; x2 = 0; Row 9 - x1 = 1; x2 =1 ;

Fig.4.11. Influence of scalo position on thread breakage

The analysis of curves (Figs. 4.1- 4.3) plotted by the obtained equations shows that the variation of **Y** from x1 and x2 has the form of concave parabolas (Figs. 4.9, 4.10). The influence of x3 (position of the rock relative to the breast of the loom Fig. 4.11) on **Y** is represented by a convex parabola with minimum y values at x_3 equal to - 1.68 and + 1.68 respectively. Consequently, when producing this fabric, it is advisable to raise the scala as much as possible, or to lower it as much as possible with respect to the sternum, and raising the scala (x_3=+1.68), as shown in Fig. 4.3 leads to the lowest breakage. Separately plotted curve of the change of breakage **Y** from the position of the offset x2 at zero value x1 (tension of the main threads) and the maximum raised rock x_3=+1,68 shows (see Fig. 4.10) that at x2=0 it is possible to reduce the breakage of threads in 2 times, ie the parameters will have the following values: tension of the main threads - 20 cN (per 1 thread); the value of the offset - 15 mm.; the position of the rock relative to the sternum - (+25) mm. At these values of parameters breakage of main threads will not exceed 0,05 breaks per 1m.

CONCLUSION

Analysis of literary sources shows that a number of scientific works on the study and design of fabric structure have been carried out. The works are mainly oriented to the study of fabric structure, thus designing of fabrics is carried out by a certain thickness, by fabric filling, strength, phase order of the woven structure, linear densities of yarns. The structure of clothing fabrics and their design are insufficiently studied, in particular, the design of clothing fabrics according to a given air permeability is of topical importance. At manufacture of suit fabrics the influence of drapery factors on the fabric structure is insufficiently studied. The criterion of fabric structure evaluation is air permeability, which determines quantitative and qualitative evaluation of fabric structure. The technique of designing clothing fabrics according to the given air permeability indicators, in particular, the air permeability coefficient, the percentage of the fabric area not filled with fibre and filled, the average number of threads, the fabric density, as well as the equation of air permeability for a given fabric is developed. The difference between the experimental values of air permeability and the equation of calculated values for a given fabric was 16% depending on the weave of the fabric, yarn composition, type and twist. The equation of the air permeability coefficient of square weave fabric, which takes into account the sum of the numerator and denominator of the main weaves, is proposed. A study of the parameters of square weave fabric as a function of shear, air permeability and surface density has been carried out. It is found that the average thread count and surface density decrease with increasing thread count and air permeability in the square structure fabric. It is characteristic that the values of parameters such as air permeability coefficient, percentage of uncovered fabric area and relative fabric density remain constant for air permeability. A method for analysing the parameters of square structure fabrics has been developed on experimental fabric samples. The relative density of fabric with fabric parameters of square structure fabric cut fabric, percentage of uncovered area of fabric, air permeability coefficient, pressure level index and air permeability were determined. Relative density of fabric with fabric parameters with square fabric cut structure, percentage of uncovered fabric area, air permeability coefficient, pressure level index and air permeability were determined. In the metric system, increasing the degree of yarn twist increases the percentage of uncovered fabric area, air permeability coefficient and breathability of square structure fabric while decreasing the relative density and pressure rating of the fabric. The tension of 1/1 warp (short cover) square structure fabric on the loom is higher than that of 1/2, 1/3 and 1/4 warp (long cover) fabrics. The surface density, fill factor (SF) and tie-back coefficient (TBC) increase with increasing number of yarn turns in square structure fabrics. Comparative studies based on structural properties and physical and mechanical properties of standard and developed suit fabrics were carried out. An alternative combination of fibre blends for the suit fabric, i.e. 90% artificial bamboo and 10% wool fibre, was founded. The new suit fabric was shown to have superior surface density, breathability, abrasion resistance, tensile strength, stretch and moisture absorption.

LITERATURE

1. Sabit Adanur. HANDBOOK OF WEAVING. CRSPRESS. Boca Raton London New York Washington, DC.-2001.-P.-488.
2. T.A.Ochilov, M.^ulmetov, S.A.Khamraeva va b. "Tukimachilik materialshunosligi" Toshkent: Adabiyotlar uchkunlari, 2018.-310b.
3. Khamrayeva S.A. Increase of wear resistance of fabrics by optimisation of parameters of their formation. Dissertation ... D. Sc. - T. - 2010.
4. Daminov A.D. Fundamentals of structure forecasting and design of textile fabrics. Dissertation. On sosis. sch. st. of doctor of technical sciences. T./ 2003g-300str.
5. Lukmonov H. N., Alimbayev E. Sh. Improving the quality of silk fabrics produced on STB machines // J. Silk. 1991. - №5. -Б. 25-27.
6. Hikmatullaeva M.R. Development of technology of silk-cotton fabrics assortment production: Cand. Sci. (Techn.) Dissertation. - T.: TITLP, 2000. - 15 б.
7. Abramova I. A. Development of design method of technological process of weaving: Author's disc. of technical sciences. - M. MSTU, 2004. - 16 б.
8. Khurram Shehzad Akhtar, Ali Afzal, Kashif Iqbal, Zahid Sarwar, Sheraz Ahmad. Investigation of Manufacturing and Processing Techniques on Shade Variation and Performance Characteristics of Woven Fabrics. // Journal of Engineered Fibers and Fabrics. Volume 13, Issue 3 - 2018. 71-77 p.
9. Kulabusheva I. V. Development of the method of designing parameters of structure and technology of fabrics manufacturing: Cand. Sci. (Techn. Sci.). - MOSCOW STATE TECHNICAL UNIVERSITY, 2003. - 15 б.
10. Masudur Rahman ANM. Influence of stitch length and structure on selected mechanical properties of single jersey knitted fabrics with varying cotton percentage in the yarn. Journal of Textile Engineering & Fashion Technology. Volume 4 Issue 2 - 2018. 189 p.p.
11. Nikishin V. B. Development of the automated method of calculation of the fabric structure parameters: Cand. B. Development of the automated method of calculation of parameters of tissue structure: Author's disc. of technical sciences. - MOSCOW STATE TECHNICAL UNIVERSITY, 2002. - 16 б.
12. Ziyatdinova V. V. Development of optimum technological parameters for manufacturing of high-density fabrics on STB shuttleless weaving machines: Cand. Sci. (Techn. Sci. Dissertation). - M.: MGTU, 1995. - 15 б.
13. Beskhlebnaya S. E. Development of the calculation method for the

volume of through pores in fabrics of main and derivative weaves. Author's thesis ... Cand. of Technical Sciences. - MOSCOW STATE TECHNICAL UNIVERSITY, 2004. - 16 б.
14. Bakaev M.H. Research and improvement of the technological process of tempering and warp tension in the production of fabrics from natural silk: Cand. of Sci . Techn. - T.: TITLP, 1993. - 144 б.
15. Kamolova S., Islomova F., Alimbayev E.S. Gazlamalarda tanda va arkok iplarining siljishga bardoshlik khusususiyatini oshirish // J. Ipak. -2001.-№2. - Б. 25-27.
16. Mogilny A. N. Development of technology, design methods and research of structure and properties of textile materials for technical purposes: Ph. D. in Technical Sciences. - St. Petersburg: SPGUTiD, 2000. - 36 б.
17. Onikov E.A., Khagaueva S.A. Cotton fabrics with higher resistance to abrazion interrnatio- nal textile reports //Melliand Textilberichte. -Germany, 2002.-No.1.-C.37-38.
18. Federenko N.A. Estimation of rationality of fabric structure // Textile industry. 1997.-№2. 44
19. Yunuskhodjaeva M. R. Yangi tarkhibli kotirma matolar tekhnologi va ta\lili: Dis. tekhnika fanlari nomzodi. - T.: TTESI, 2001. - 142 б.
20. Mikhailyuk O.Yu., Onikov E.A. Analysis and choice of formula for determination of assortment possibilities of the machine // J. Textile industry. - 2003.- №1-2. -Б 18-19.
21. Sukhova L.V. Method of designing fabrics with shadow transitions of shaped weft yarns. Dissertation of Candidate of Technical Sciences - M. - 1998.
22. Kareva T.Yu. Development of method, technology of fabrics manufacturing of new structures and research of their structure Diss.... D. Sc. - M. - 2005.
23. C. V. Smirnov, Multivariate nomograms and fabrics, UMN, 1963, vol. 18, issue 3(111), 239-240
24. Novikov N.G. About the structure of the fabric and its design by means of the geometrical method // J. Textile Industry. - 1946. - № 2 C-54 25. Nazarova, M.V.; Romanov, V.Yu. Development of the optimum technological parameters for the production of a quill fabric with the maximum air permeability. // Journal International Journal of Applied and Fundamental Research. M.: - 2016. - No. 12 (part 3) - P. 422-425.
26. Goryachev M.V. Development of the method of estimation and calculation of air permeability of fabrics made of monofilaments. Abstract,

dissertation ... Candidate of Technical Science -M -2002. -19 c.
27. Arkhangelsky N.A. Air permeability of fabrics depending on their structure// Scientific Works. Plekhanov Institute of National Economy, 1959, 142 p.
28. Eremina N.S. Study of regularity of change of physical-mechanical and hygienic properties of fabric from its structure. M., 2002.
29. Bubentsov L.V. Influence of density on base and weft and weave on static electricity charge potential. Izv. VUZov. Technology of textile industry. 1978 № 11. c38-40.
30. Semak, Z.N. Influence of structure of multifibre dress-suit fabrics on their dielectric properties // Izv. of high schools. Technology of textile fabrics. -1981. -№ 5. - C. 36-39.
31. Martynova A.A., Slostina G.L., Vlasova P.A. Structure and design of fabrics. M., RIO MSTA, -1999.-434 pp.
32. Rakitskikh V.V. Influence of polyester fabrics weave on their recovery after buckling and stiffness value. M.: Nauka, 1994. -98c.
33. Alieva D.G. Designing a new assortment of fabric // J. Problems of textile. -Tashkent. - 2008. -№ 2. -C 30 - 34.
34. Kurdenkova A.V. Development of forecasting methods of physicomechanical properties of cotton fabrics after the action of various wear factors. Abstract, dissertation ... D. Sc. - M. 2006. 38 c.
35. Nazarova, M.V.; Fefelova, T.L. Development of the automated method of designing a fabric for overalls by thickness and surface porosity of a fabric (in Russian) // J. Modern problems of science and education. - 2007. - № 4 - C. 104-110.
36. Damyanov G.B., Bachev C.Z. Fabric structure and modern methods of its design. -M. - 1984.
37. Eremina N.S. Study of the regularity of bending waves of main and weft yarns in plain weave fabrics / / Textile industry. - Moscow, 1993.- №3.-S.36-38.
38. Efremov D.E., Amarzhargalen T. Use of parabola in geometry of fabric element// Izv. of high schools. Technol. of textile industry. 1989.-№5.-C.47-49.
39. Zhuraev A.T. Development of structures and technology of production of multilayer fabrics for footwear and clothing purposes. Abstract, dissertation of Candidate of Technical Sciences - St. Petersburg. - 1994.
40. Vasilchikova N.V. Designing, structure and properties of melange fabrics from lavsano-viscose yarn: Cand. Candidate of Technical Sciences.

Leningrad, 1968 41. Kemp A.Ah. Extension of Pierces cloth geometry to the treatment of non - circular threads " The journal of the Textile Institute" 55/1997. -t. 66-70.
42. Rachenkov O.M. Development of the calculation method of the rational parameters of the structure of the fabrics of different weave taking into account the technology of their manufacturing: Dissertation .... Candidate of Technical Sciences. - M. MTI.2000.-136 p.
43. Kozmich D.I., Dianich M.M. Investigation of air permeability of linen-lavsan fabrics for costume dresses. Technology of textile industry. Izv. vuzov, 1970, No. 5 p. 14-17.
44. GOST 12088-77 Interstate standard textile materials and textile products method of determination of air permeability
GOST 28000 -2004 Interstate Standard Moscow
45. Clothing fabrics pure woollen, woollen and semi-woollen General technical conditions. Moscow
46. Aripjanova D.U., Habibullaev D.A., Tuichiev I.I. Development of the structural scheme of formation of rational assortment in the system "kit" from woollen and polycomponent fabrics. Vestnik nauki i obrazovanie nauchno-methodicheskogo zhurnal #13 (49) october 2018. 134 pp.
47. Scherbakov V.P. Applied and structural mechanics of fibre materials. TISO PRINT, Moscow, 2013, pp.175-180
48. R. I. Orazbaeva, D.N Kodirova, S.S Rakhimkhodzhaev, A.B Djoldasova "Using the environmental properties of air permeability for designing of clothing fabrics with specified properties" IOP Conference Series: Earth and Environmental Science. AGRITECH-VI-2021 IOP Conf. Series: Earth and Environmental Science **981**(2022) 022035 doi:10.1088/1755-1315/981/2/022035.
49. R.I.Orazbayeva, D.N.Kadirova, S.S.Rakhimkhodjaev, L.A.Tureniyazova "Influence of yarn twist on the parameters of fabrics of square structure" "Modern innovative technologies in light industry: problems and solutions" materials of the international scientific-practical conference Bukhara 19-20 November 2021. 60-64 б.
50. N.B. Yusupova, S.A. Khamrayeva, R. Begmanov . Technological possibilities of obtaining fabric pattern in remiz weaving // "Fan, ta'alim va ishlab chikarish integratsilashuvi sharoitida innovatsion tekhnologii dolzarb muammolari" Respublika ilmiy - amaliy anjumani, -Toshkent, 2016, - B.159-161.
51. R.I.Orazbayeva "Innovative suit fabric" "Modern innovative technologies

in light industry: problems and solutions" materials of the international scientific-practical conference Bukhara 19-20 November 2021 228-231b.
52. Orazbayeva R.I. Research of clothing fabrics breathability of the Main // KorakαlpoFiston Respubliki oliy ta'lim muassasalari olimlarining ilmiy tuplami KDU Nukus. No. 4, 2021., 92-97 b. (05.00.00;№27)
53. Orazbayeva R.I., Kodirova D.N., Rakhimkhodjaev S.S., Turmanov I., Tureniyazova L.A. Kostyumbop tukimalarda ip buramlarining khavo utkazuvchanligiga ta'asiri // Ilim ha'm ja'miyet NDPI Journal Nukus. No. 4, 2021., 30-32 b. (05.00.00;№37)
54. Orazbayeva.R.I., D.N.Kadirova, A.B.Zholdasova. "Kiyimbop tukimalarning tuzili shva physic-mechanical course kursatkichlari tadkiki" "KorakolpoFiston Respublikasida ishlab chikarish sanoat sohalari rivozhining dolzarb muammolari" mavzusidagi Respublika ilmiy-amaliy anjuman Nukus. 26.04.2021. 84-86 б.
55. Orazbayeva R.I., Kadirova D.N., Rakhimkhodzhaev S.S., Tureniyazova L.A., Djoldasova A.B. Innovative fabric for suit // Koratsalpogaston Respubliki oliy ta'lim muassasalari olimlarining ilmiy tuplami KDU Nukus. No. 3, 2021., 167-170 pp. (05.00.00;№27)
56. R. I. Orazbaeva, L.Toreniyazova, D.Kadirova, R.Rakhimkhodzhaev "Investigation of the twist of a yarn with a square structure" Karakalpak Scientific Journal: Vol. 4: Iss. 2, Article 3. https://uzjournals.edu.uz/karsu/vol4/iss2/3 6-302021. Volume 4.
57. Orazbayeva.R.I., Tureniyazova, Sh.Musirov "Tarkibi turli hil tolalar aralashmasining matolardagi sifat kursatkichlari" "KorakolpoFiston Respublikasida ishlab chikarish sanoat sohalari rivozhining dolzarb muammolari" mavzusidagi Respublika ilmiy-amaliy anjuman Nukus. 26.04.2021. 91-93 б.
58. Orazbayeva.R.I., D.N.Kadirova "Design of garment fabrics on breathability" 6 TH MANCHESTR, ENGLAND CONFERENCE-2022. September 25 W.pp.
59. R.I. Orazbayeva, A.B. Joldasova. Joldasova., A.T. Orazbaeva. Orazbaeva. Changes in the full composition of the physical and mechanical characteristics of costumbop fabrics // International Journal of Innovations in Engineering Research and Technology [IJIERT] ISSN: 2394-3696, Website: ijiert.org, August 14th, 2020. 102-10 p.p. https://repo.ijiert.org/index.php/ijiert/article/view/223. 102-107 p.p.
60. S.A.Hamraeva, E.T.Laysheva, Z.F.Valiyeva. Mahalliy jun tolasining geometri xususiyatlarini standart usuli va akustik qurilmasi yordamida

aniqlash. Toshkent - 2001.Tukmachchlik mummolari. №3. 44-46 б.
61. Yildiray Turhan, Recep Eren. The effect of loom settings on weavability limits on air-jet weaving machines. // Textile Research Journal Volume: 2018y. 60 issue:7, page(s): 389-404.
62. Del R.A., Afanasiev R.F., Chubarova Z.S. Hygiene of clothes: Textbook for universities. - 2nd edition, revised and supplemented - M.: Legkombytizdat, 1991. - 160 c.
63. Sayfieva M.A. Tukuv usulida badiyiy bezash asosida yangi tarkibli gazlamalar yaratish. Diss....techn. fan.nom. -Toshkent: TTESI. 2009.-1126.
64. N.B.Yusupova, N.R.Sodikova . Costumebop matolar assortment assortmentlik khusususiyatlari tahlili // "Fan, talim, ishlab chikarish integratsiyalashuvi sharoitida pakhta tozalash, tukimachilik, yengil sanoat, matbaa ishlab chikarish innovative technology dolzarb muammolari va ularning echimi" Respublika ilmiy - amaliy anjumani. -Toshkent, 2019, - B.87-90
65. Sklyannikov V.P., Afanasiev R.F., Mashkov E.N. Hygienic assessment of materials for clothing. - Moscow: Legkombytizdat, 1985. - 144 c.
66. Boimuratov B.H. Improvement of warp tempering and tensioning process on shuttleless weaving machines. Dissertation of Candidate of Technical Science - Tashkent. 2001,136 б.
67. Shumkorova Sh. P., Yuldasheva M. T., Yadgarova H. I., Begmanov R. A., Valieva Z. Influence of fibre composition on physical and mechanical properties of suit fabrics // Young Scientist. Moscow. 2014. - №9. -C. 235-238.
68. Martynova A.A., Starostina G.L., Vlasova N.A. Fabric structure and design: Textbook for universities. - Moscow: MSTU, (International Education Programme), 1999.
69. Arkhangelsky N.A. Air permeability of tissues depending on their structure// Scientific Works. Plekhanov Institute of National Economy, 1959.
70. Arkhangelsky N.A. et al. Performance properties of fabrics and modern methods of their evaluation. Rostekhizdat. Moscow. 1960. 475 pp.
71. N. Gokarneshan. Fabric Structure And Design. New Age International Publishers. 2004 -152p.
72. Kondratsky E.V. Dependence of air permeability of fabrics of different structure on pressure drop. Abstract of dissertation of Candidate of Technical Sciences, M.: MTI, 1972.
73. Beskhelebnaya S.E. Development of calculation of through pores in fabrics of main and derivative weaves: Cand. Sci. (Techn.) Dissertation. - M.,

2004.
74. Guseinova T. S. Commodity management of sewing and knitted goods: Textbook for universities. - Moscow: Economics, 1991. - 287 c
75. Zhikharev A.P. Material science in the production of light industry products. M, 2005
76. Rakhimkhodjaev S.S., Kadyrova D.N. Theory of tissue structure. Textbook. Tashkent. Adabiyot uchkunlari. 2018. - 212 pp.
77. Goryachev M. V. Development of the method of estimation and calculation of air permeability of fabrics made of monofilaments. Abstract of dissertation of the candidate of technical sciences. M.-2002.
78. Zhernitsin Y.L., Gulamov A.E. Methodical instruction on performance of research and laboratory works on testing of textile products. Tashkent. 2007. 96 pp.
79. Kukin G.N. et al. Textile material science (textile fabrics and products). M., Legpromizdat, 1992
80. Amzayev L.A., Jumaniyozov Q.J, S.L.Matismailov. Tadqiqot uslub va vositalari. Darslik. Toshkent.: G'.G'ulom. 2014-192bet.
81. Sevostyanov A.G. Methods and means of research of mechanic-technological processes of textile industry.-M.: Legkaya Industriya, 1980. - 392 c.
82. B.K. Behera and P.K. Hari. Woven Textile Structure. Woodhead Publishing Series in Textiles. 2010y. 450p.
83. Raximxodjaev S.S , D.N.Qodirova To'qima loyialashning zamonaviy usullari. Darslik.-T.: Adabiyot uchqunlari. 2018-144b
84. D.T.Nazarova, N.B.Yusupova, N.Pulatov**.** Tukuv dastgohida sifatli tukima khosil bulishida tanda va arkok iplari tarangligining ahamiyati // "Pahta tozalash, tukimachilik, yengil sanoat, matbaa ishlab chikarish techno-technologiclarni modernisationlash sharoitida iktidorli yoshlarni innovatsionnogo foyalari va ishlanmalari" Respublika ilmiy amaliy anjumani. - Toshkent, 2018, -B.83-85.
85. R.I.Orazbayeva., Kadirova D., Zholdasova A "Analysis of assortments of clothing fabrics" Innovative approaches in modern science *Collection of articles on the materials of LXXIX international scientific-practical conference. Moscow* October 2020, № 19 (79) 107-111 b.
86. N.B. Yusupova, R.I. Orazbayeva, M.T. Shamuratov Studies of the influence of weaving machine dressing parameters on fabric properties // Tukimachilik sanoati techno-technologi va innovatsiyalar asosida ularni rivozhlantirish (Academician M. Khozhinovaning 105 yilligiga baF.).-

Toshkent, 2017,-B.397-399.
87. Orazbayeva R.I. Air permeability of clothing fabrics of main weaves // Ilim ha'm ja'miyet NDPI Journal Nukus. No. 4, 2021., 32-34 p. (05.00.00;№37)
88. Onikov E.A. Designing of weaving mills. M, Inform-Znanie, 2005, p.432.
89. https://izvolokna.com/materialy/tkani/naturalnye/rastitelnye/bambukovoe-volokno.html
90. Kavokin S.G. et al. "Reference book on wool quality" part II, Moscow, L.I., 1975, pp. 388-389.
91. R.I.Orazbayeva, S.S.Rakhimkhodjaev, D.N.Kadirova, A.B.Zholdasova, L.A.Tureniyazova, D.A.Rajapova Patent. Costume fabric Uzbekistan Respublika Respubliki Adliya vazirligi huzuridagi Intellectual mulk agenligi. NO. IAP 06636. 08.11.2021 й
92. Zhernitsin Y.L., Gulamov A.E. Performance of research and laboratory works on testing of textile products. M.U. Tashkent, 2007. 88 pp.
93. Orazbayeva.R.I., A.B.Zholdasova, G.Baimuratova. Kuylaklik matolaining sifat kursatkichlariga tolalar tarkibining tasiri "Korakozpogiston Respublika ishlab chikarish sanoat sohalari rivozhinning dolzarb muammolari" Respublika ilmiy-amaliy anjuman. Nukus. 26.04.2021. 86-88 б.
94. UZ patent FAP 01488. 16.03.2020. Zhun tolasini tekislovci va yumshatuvci kurilma. Khamraeva S., Orazbayeva R., Khudaiberganova Z., Yusupova N.B., Ubaidullayeva B., Toshev A., Valiyeva Z.

Printed by Books on Demand GmbH, Norderstedt / Germany